SMALL-SCALE WIND POWER

SMALL-SCALE WIND POWER

DESIGN, ANALYSIS, AND ENVIRONMENTAL IMPACTS

JOHN ABRAHAM AND BRIAN PLOURDE

# MOMENTUM PRESS

MOMENTUM PRESS, LLC, NEW YORK

Small-Scale Wind Power: Design, Analysis, and Environmental Impacts
Copyright © Momentum Press®, LLC, 2014.

All rights reserved. No part of this publication may be reproduced, stored in a retrieval system, or transmitted in any form or by any means—electronic, mechanical, photocopy, recording, or any other—except for brief quotations, not to exceed 400 words, without the prior permission of the publisher.

First published by Momentum Press®, LLC
222 East 46th Street, New York, NY 10017
www.momentumpress.net

ISBN-13: 978-1-60650-484-0 (print)
ISBN-13: 978-1-60650-485-7 (e-book)

Momentum Press Environmental Engineering Collection

DOI: 10.5643/9781606504857

Cover design by Jonathan Pennell
Interior design by Exeter Premedia Services Private Ltd.,
Chennai, India

10 9 8 7 6 5 4 3 2 1

Printed in the United States of America

ABSTRACT

In today's world, clean and robust energy sources are being sought to provide power to residences, commercial operations, and manufacturing enterprises. Among the most appealing energy sources is wind power—with its high reliability and low environmental impact.

Wind power's rapid penetration into markets throughout the world has taken many forms. In some cases, wind power is produced in large industrial wind farms by large horizontal-axis wind turbines (HAWTs) that supply power directly to the grid. In other cases, wind power is produced more locally, at or near the site of power usage. In these cases, the wind is typically generated by smaller wind turbines that are in single-unit installations or in clusters of units. Regardless of the manifestation of the wind turbine systems, much thought must be given to the selection of the turbine, the economic viability of the installation, and the integration of the system to the connecting grid.

It is the intention of this text to discuss these issues in detail so that appropriate decisions can be made with respect to wind power design, testing, installation, and analysis. The specific focus is on small-scale wind systems. While there is no universal definition of small-scale wind power, it generally refers to systems that produce only a few kilowatts, can be installed in constrained spaces, and have a small footprint.

The text is written for a wide range of audiences and much of the discussion deals with the design of various small-wind systems, including horizontal-axis wind turbines (HAWTs) and vertical-axis wind turbines (VAWTs). Both of which will be discussed in detail.

The design of wind turbines takes advantage of many avenues of investigation, all of which are included here. Analytical methods that have been developed over the past few decades are major methods used for design. Alternatively, experimentation (typically using scaled models in wind tunnels) and numerical simulation (using modern computational fluid dynamic software) are also used and will be dealt with in depth in later chapters.

In addition to the analysis of wind turbine performance, it is important for users to assess the economic benefits of using wind power, therefore, an entire chapter of this book is devoted to this topic. Finally, the decision to utilize wind power can take advantage of past experience gained through case studies. Such case studies help elucidate the issues that users must consider, from siting and mechanical complications, to performance and maintenance. Experience gained from case studies will be showcased here.

This text is intended to be useful to engineers, scientists, wind-power users, installers, and investors. It is written in part as a hands-on guide and a development tool. The authors are leading contributors from around the world, collectively, they represent the present state-of-the-art with respect to small wind power.

KEYWORDS

clean energy, Darrieus wind turbines, horizontal-axis wind turbines, local power production, off-grid energy generation, remote wind power, renewable energy, Savonius wind turbines, small-scale wind power, sustainable energy, turbine blades, turbine rotors, urban wind generation, vertical-axis wind turbine, wind turbine aerodynamics, wind turbine rotor design, wind turbines

CONTENTS

LIST OF FIGURES — vii

LIST OF CONTRIBUTORS — ix

CHAPTER 1 INTRODUCTION TO SMALL-SCALE WIND POWER — 1

CHAPTER 2 FINANCIAL AND IMPLEMENTATION CONSIDERATIONS OF SMALL-SCALE WIND TURBINES — 17

CHAPTER 3 DESIGN OF DARRIEUS-STYLE WIND TURBINES — 45

CHAPTER 4 DESIGN OF SAVONIUS-STYLE WIND TURBINES — 65

CHAPTER 5 DESIGN OF HORIZONTAL-AXIS WIND TURBINES — 93

CHAPTER 6 NUMERICAL SIMULATIONS OF SMALL WIND TURBINES—HAWT STYLE — 129

CHAPTER 7 CASE STUDIES OF SMALL WIND APPLICATIONS — 147

INDEX — 173

LIST OF FIGURES

Figure 1.1.	A small-scale VAWT wind-power system attached to a cellular communication tower.	4
Figure 1.2.	Close-up photograph showing a small-scale turbine with support brackets and positioning of electrical generator.	5
Figure 1.3.	Photograph of an array of vertical-axis Darrieus-style turbines.	5
Figure 1.4.	Close-up photograph of the rotors of a Darrieus-style VAWT.	6
Figure 1.5.	Close-up photograph of a propeller HAWT.	7
Figure 1.6.	Rotating HAWT rotor with contoured blades.	8
Figure 1.7.	Photograph of a Darrieus VAWT connected to a solar panel, positioned in a constrained urban location.	10
Figure 1.8.	Annual average wind speeds at 30 m height across the United States.	12
Figure 1.9.	Annual average wind speeds at 80 m height across the United States.	12
Figure 1.10.	Annual average offshore wind speeds in regions surrounding the United States.	13
Figure 1.11.	Average wind speeds across China, 70 m height.	14
Figure 2.1.	A computer rendering of a small-scale Savonius VAWT rotor.	19
Figure 2.2.	Images of small-scale HAWT.	21
Figure 2.3.	Small-scale VAWT on Lincoln Financial Field.	22
Figure 2.4.	Power Curve for small-scale wind turbine with a sweet area of 5 m^2 and a C_p of 0.2.	24

LIST OF FIGURES

Figure 2.5. A small-scale wind turbine designed for telecommunication base transceiver stations during installation on U.S. Department of Transportation telecommunication tower.   27

Figure 2.6. Power curve for small-scale wind turbine with a swept area of 6 m² and C_p of 0.3.   36

Figure 2.7. Average wind speed map of Kenya.   37

Figure 3.1. Darrieus Rotor Near Heroldstatt in Germany.   46

Figure 3.2. Straight-bladed Darrieus VAWT (SB-VAWT).   47

Figure 3.3. Flow velocities of SB-VAWT.   55

Figure 3.4. Force diagram for a blade airfoil.   56

Figure 4.1. Two-bladed "S"-shaped and helical-shaped Savonius-style vertical-axis wind turbines.   66

Figure 4.2. Dimensions of a Savonius-style vertical-axis wind turbines.   67

Figure 4.3. Historic Gold Mine Savonius-Darrieus combined small-scale vertical-axis turbine in Taiwan (November, 2009).   68

Figure 4.4. Various flow patterns around a Savonius-style wind turbine, (I: Free stream flow, II: Coanda type flow, III: Dragging-type flow, IV: Overlapping flow, V: Separation flow, VI: Stagnation flow, VII: Returning flow, VIII: Vortex flow).   70

Figure 4.5. Power coefficient of Savonius-style "S"-shaped wind turbines along with other counterparts.   71

Figure 4.6. Torque coefficient of Savonius-style "S" shaped wind turbines along with other counterparts.   71

Figure 4.7. Savonius-style turbines.   74

Figure 4.8. Darrieus-style turbines.   74

Figure 4.9. Examples of Savonius-style turbines with different aspect ratios and overlap ratios.   75

Figure 4.10. Effect of overlap ratio on the performance of Savonius-style wind turbines.   76

Figure 4.11. Application of end plates and multi-staging in to the design of Savonius-style wind turbines.   77

Figure 4.12. Effect of multi-staging on the torque performance of Savonius-style wind turbine.   78

Figure 4.13.	Comparative analysis of two and three-bladed Savonius-style turbines.	79
Figure 4.14.	Demonstration of most common Savonius-style blade profiles patented by Savonius and Benesh.	80
Figure 4.15.	Optimized blade profile of Savonius-style turbine	80
Figure 4.16.	Application of Savonius style wind turbines in communication towers with a modified blade profile.	81
Figure 4.17.	A twisted bladed Savonius-style vertical-axis wind turbine "Gale" by Tangarie.	82
Figure 4.18.	Photograph of a 3 kW helical vertical-axis wind turbine	83
Figure 4.19.	Close-up photograph showing testing of conventional Savonius-style turbines in a wind tunnel with open test section facility.	83
Figure 4.20.	Testing of Savonius-style helical wind turbines in a wind tunnel with a closed test section.	84
Figure 4.21.	Effect of blockage on the performance of Savonius-style wind turbine.	85
Figure 4.22.	Design and dimensions of a Savonius-style wind turbine.	88
Figure 5.1.	Rotor power coefficient as a function of tip speed ratio for various rotor configurations	94
Figure 5.2.	Components of a small HAWT.	96
Figure 5.3.	Schematic of a teetering mechanism.	97
Figure 5.4.	Small HAWT installed on the rooftop of the Western University Engineering building.	98
Figure 5.5.	Sky Serpent design installed in Tehachapi, California	99
Figure 5.6.	Typical power curve of small HAWTs.	100
Figure 5.7.	Wind turbine rotor modeled by an actuator disc.	103
Figure 5.8.	Schematic of the rotating actuator disc with an annular ring of radius r.	106
Figure 5.9.	Tangential velocity changes across the rotor disc.	107
Figure 5.10.	Theoretical maximum power coefficient as a function of tip speed ratio for an ideal HAWT with and without wake rotation.	108
Figure 5.11.	Blade geometry for blade element analysis.	109
Figure 5.12.	Blade element geometry, velocities, and forces.	110

Figure 5.13.	BEM prediction of power output compared with wind tunnel measurements.	113
Figure 5.14.	Comparison between experimental results and BEM theory prediction with and without tip loss corrections.	115
Figure 5.15.	Effect of stall delay correction on the rotor performance prediction by BEM.	116
Figure 5.16.	Lift and drag coefficients for FX 63 137 airfoil obtained through wind tunnel testing and JavaFoil.	119
Figure 5.17.	Schematic drawing of a vehicle-based wind turbine prototype testing setup.	121
Figure 5.18.	Wind tunnel test setup of a small HAWT.	122
Figure 5.19.	Power coefficient variation with tip speed ratio for a three-bladed HAWT.	123
Figure 5.20.	Comparison of experimental and theoretical power coefficients.	124
Figure 5.21.	The theoretical and experimental power curve of the rotor for a wide range of wind speeds.	125
Figure 5.22.	Percentage of difference between predicted power by BEM theory and the measured power.	126
Figure 6.1.	Streamline patterns from a small horizontal axis wind turbine by CFD simulation.	135
Figure 6.2.	Schematic of wind profile.	135
Figure 6.3.	Meshing for wind turbine blade and rotor CFD simulation.	137
Figure 6.4.	Meshing for entire wind turbine system CFD simulation.	138
Figure 6.5.	Pressure coefficient (C_p) on the blade surface along the entire blade length.	143
Figure 6.6.	Power curves for typical 10 kW HAWT.	143
Figure 7.1.	View of the Rasoon camp.	149
Figure 7.2.	View of the 2 kW wind turbine.	150
Figure 7.3.	The connection diagram of the 2 kW wind system.	150
Figure 7.4.	Electrical wind pump.	153
Figure 7.5.	Naima project equipment layout.	154
Figure 7.6.	Wind turbine installed at Heelat Ar Rakah.	156

Figure 7.7	The 65 kW wind turbine installed in Appalachia, Wales, Alaska.	157
Figure 7.8.	The 900 W wind turbine on hillside above Douglas County home.	160
Figure 7.9	Monthly average wind speed at Alasfar site.	161
Figure 7.10	Monthly power consumption of Alasfar site.	162
Figure 7.11.	PLC Module and touch screen for easy turbine control.	162
Figure 7.12.	Free standing tower for 20 kW wind turbine.	163
Figure 7.13.	Power curve of 20 kW wind turbine installed at Alasfar site.	165
Figure 7.14.	Probability distribution function for the wind at the considered site.	166
Figure 7.15.	A view of Al-Ibrahimiyah wind farm.	168
Figure 7.16.	Small wind farm near the community of Ramea, Canada.	169

List of Contributors

J.P. Abraham, University of St. Thomas, School of Engineering, St. Paul, MN, USA

M.H. Alzoubi, Yarmouk University, Irbid, Jordan

H. Hangan, Western University, WindEEE Research Institute, London, Ontario, Canada

M. Islam, Department of Mechanical and Manufacturing Engineering, Schulich School of Engineering, University of Calgary, Alberta, Canada

W.J. Minkowycz, University of Illinois, Chicago, Department of Mechanical and Industrial Engineering, Chicago, IL, USA

P.O. Okaka, Kenyatta University, Mechanical Engineering Department, Nairobi, Kenya

B.D. Plourde, University of St. Thomas, School of Engineering, St. Paul, MN, USA

M. Refan, Western University, WindEEE Research Institute, London, Ontario, Canada

S. Roy, Indian Institute of Technology Guwahati, Department of Mechanical Engineering, Guwahati–781039, India

U.K. Saha, Indian Institute of Technology Guwahati, Department of Mechanical Engineering, Guwahati–781039, India

E.D. Taylor, University of St. Thomas, School of Engineering, St. Paul, MN, USA

J.C.K. Tong, University of Minnesota, MN, USA

CHAPTER 1

INTRODUCTION TO SMALL-SCALE WIND POWER

B.D. Plourde, E.D. Taylor, W.J. Minkowycz, and J.P. Abraham

In this opening chapter, the authors introduce the topic of small-scale wind power by discussing its power production characteristics and its applications. The authors outline the advantages of wind power over other energy sources and the specific situations that are uniquely suited for small-scale wind solutions. Furthermore, a discussion of the available wind speed information is presented and a cursory discussion of financial issues is given. Chapter one provides readers with outlines of subsequent chapters and also serves as a guide for the book in its entirety.

1.1 INTRODUCTION

Wind power's rapid penetration into markets throughout the world has taken many forms. In some cases, wind power is produced in large industrial wind farms by large horizontal-axis wind turbines (HAWTs) that supply power directly to the grid. In other cases, wind power is produced more locally, at or near the site of power usage. In these cases, the wind is typically generated by smaller wind turbines that are in single-unit installations or in clusters of units. Regardless of the manifestation of the wind turbine systems, much thought must be given to the selection of the turbine, the economic viability of the installation, and the integration of the system to the connecting grid.

It is the intention of this text to discuss these issues in detail so that appropriate decisions can be made with respect to wind-power design, testing, installation, and analysis. The specific focus is on small-scale wind systems. While there is no universal definition of small-scale wind power, it generally refers to systems that produce up to a few kilowatts and can be installed in constrained spaces and with a small footprint.

The text is written to be of use to a wide range of audiences. Much of the discussion contained herein deals with the design of various small-wind systems including HAWTs and vertical-axis wind turbines (VAWTs). Among the vertical-axis variants are two major classifications: Darrieus and Savonius-style rotors. The former is often colloquially referred to as the *egg-beater* turbine because of its shape. The latter is often, but not always, fashioned from a cylindrical drum that is bisected with the two halves offset some distance from a common rotation axis. The Darrieus-style turbine has its rotation caused by lift forces, whereas the Savonius-style turbine is driven by drag. Both of these classes will be discussed in detail.

The design of wind turbines takes advantage of many avenues of investigation, all of which are included here. Analytical methods, which have been developed over the past few decades, are one major method used for design. Alternatively, experimentation (typically using scaled models in wind tunnels) and numerical simulation (using modern computational fluid dynamic software) are also used and will be dealt with in depth in later chapters.

In addition to the analysis of wind turbine performance, it is important for users to assess the economic benefits of using wind power. An entire chapter of this book is devoted to this topic. Finally, the decision to utilize wind power can take advantage of past experience gained through case studies. Such case studies help elucidate the issues that users must consider, from siting and mechanical complications to performance and maintenance. Experience gained from case studies will be showcased in this book.

This text is intended to be useful to engineers, scientists, wind-power users, installers, and investors. It is written partly as a hands-on guide and partly as a development tool. The authors are leading contributors from around the world. Collectively, they represent the present state-of-the-art with respect to small-wind power.

1.2 WHAT IS SMALL-SCALE WIND POWER?

As indicated earlier, small-scale wind power is wind energy that is generated at the site of utilization and is typically in the few kilowatt range. Small-scale wind turbines are often connected directly to the devices that require electricity, or more commonly, to a power charging station such as a battery array. The necessity of an energy storage system is obvious when the intermittency of wind power is recognized. It is often the case that power is available when not needed by the user, or conversely, electricity is needed but not yet generated. Storage systems help alleviate this condition by storing excess energy for later use.

Small-scale wind power is often combined with complementary energy sources. For instance, the combination of small-scale wind with solar power allows energy to be generated more consistently, when either sunlight or wind is present. Other combinations include small-scale wind power with more conventional energy sources (such as diesel generators) that can be used to guard against scenarios when more energy is required by the user than is supplied by the system.

Regardless of the actual manifestation, whether a stand-alone wind turbine or a wind turbine combined with a complementary power source, the electronic infrastructure must be given some consideration. The electronic components and systems for small-scale wind turbines have been active areas of research [1–5].

An example of a small-scale wind-power system is shown in Figure 1.1. The VAWT turbine is shown positioned to the left-hand side of a communication tower at an elevation of approximately 50 m. The turbine rotors are staged (three stages) and oriented approximately 60° from their neighbor. Staged orientation of VAWTs helps to guard against stalling, mechanical vibrations, and cyclical variations in power.

A close-up view of the wind turbine is shown in Figure 1.2. There, it is seen that the turbine in question is a drag device (Savonius-style turbine). These devices are typically large curved surfaces that are contoured to create unequal forces on their two opposing sides so that there is a net torque about the central axis. The photographs shown in Figures 1.1 and 1.2 are meant to provide some information on the physical size of small-scale turbines. Turbines within this category are typically a fraction of a meter to a few meters in dimension, depending on the specific system.

The examples illustrated in Figures 1.1 and 1.2 are stand-alone turbines. In some situations, however, turbines can be used in collections. An example of a wind turbine array is shown in Figure 1.3. In this figure, a multitude of Darrieus-style turbines can be seen, each of which rotates about a vertical axis. The study from which the figure originated was focused on developing an understanding of how disturbances in air flow created by one turbine might impact the performance of a neighbor [8,9].

Another variation of this Darrieus style is shown in Figure 1.4. There, three airfoils are clearly seen, which are able to rotate about a vertical axis. They are driven by lift forces exerted on the rotor by the wind, similar to the forces that act on an airplane wing.

Figure 1.1. A small-scale VAWT wind-power system attached to a cellular communication tower (reproduced by permission Plourde et al. [6]. © 2011 by International Frequency Sensor Association).

Some VAWTs are neither purely Savonius nor Darrieus style. For instance, cup rotors [10] and other shapes have been used, which take advantage of both drag and lift forces simultaneously. Regardless of the design, VAWTs tend to be less sensitive to the direction of wind flow than their HAWT counterparts. They can operate in constrained spaces such as on rooftops with other equipment nearby or on cellular towers with obstructions nearby. Savonius VAWTs typically begin rotating at relatively low wind speeds (3–5 m/s), generating high torque with a low rate of revolution.

INTRODUCTION TO SMALL-SCALE WIND POWER • 5

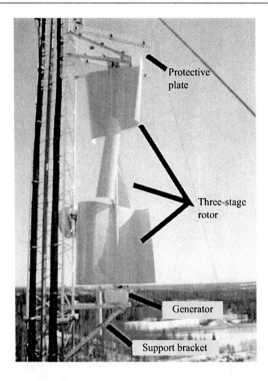

Figure 1.2. Close-up photograph showing a small-scale turbine with support brackets and positioning of an electrical generator (reproduced by permission from Abraham et al. [7]. © 2011 by AIP Publishing LLC).

Figure 1.3. Photograph of an array of vertical-axis Darrieus-style turbines. *Source*: Courtesy of John O. Dabiri/California Institute of Technology [8,9].

Figure 1.4. Close-up photograph of the rotors of a Darrieus-style VAWT.
Source: Courtesy of Aeolos Wind Energy, Ltd (http://www.windturbinestar.com/).

Despite these characteristics, the most commonly encountered small-scale wind turbine is the HAWT style. The reason for this is the higher efficiency of most HAWTs and the simplicity of installation. A photograph showing a typical small HAWT is presented in Figure 1.5. These turbines are smaller versions of the propeller-style large grid-connected turbines.

A second figure showing a small-scale HAWT with a curved rotor is provided in Figure 1.6. In some instances, contouring of blades allows more careful control of the relative velocity between the wind and the rotor at each location along the length of the rotor arm. As with large grid-level HAWTs, the shape, width, and angle of the blades should be carefully designed because the relative velocity between the wind and the rotor blade varies with the distance from the axis of rotation—details of the design methodology associated with HAWTs are given in Chapter 5.

1.3 APPLICATIONS OF SMALL-SCALE WIND POWER

As evident from the earlier discussion, small-scale wind power may take many forms and support a multitude of residential, commercial, and social activities. Perhaps, the most obvious one is the use of small-scale wind power to generate limited amounts of electrical power, typically up to a few kilowatts. Inasmuch as wind flow is not constant, the power generation is expressed as the average associated with the average wind speed at the location. It must be recognized that when wind speeds increase and decrease, the power generation rate increases and decreases as well.

This power may be used in areas where grid-based electricity is not reliable, where grid electricity is not available at all, where grid electrical costs are high, or for persons who wish to generate electricity without reliance upon grid power and the greenhouse gas emissions that naturally result from conventional electrical generation.

The small size of the wind turbine systems discussed here enables them to be mounted on relatively short supporting towers (a few 10s of meters at the most), or they can be attached to existing infrastructure (sides or roofs of buildings, cellular communication towers, etc.). As a

Figure 1.5. Close-up photograph of a propeller HAWT.
Source: Courtesy of John Hay, University Nebraska Lincoln.

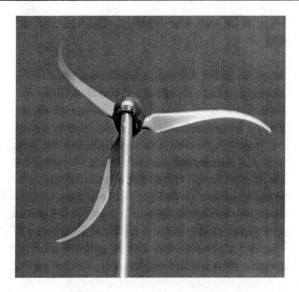

Figure 1.6. Rotating HAWT rotor with contoured blades.
Source: Courtesy of John Hay, University Nebraska Lincoln.

consequence of their positioning in urban or rural areas where spaces may be constrained and where obstructions to wind flow exist, small-scale wind systems tend to be less dependent on wind speed, turbulence levels, and wind direction than their large grid-level counterparts. These general abilities, however, do not obviate the need for proper siting because insufficient wind strength or poor wind quality can render a small-scale wind-power system economically nonviable.

Wind-power systems that are used in residential locations are able to provide significant portions of the home electrical requirements. This is particularly true if the system is accompanied by a system that allows the storage of energy when it is generated but not needed by the resident. The stored energy is then drawn down during times of electrical demand. In some geographic locations, in particular, in many U.S. states, residences are allowed to generate and sell electricity to the grid supplier. In such a situation, the wind-power system can be used to generate revenue.

For commercial uses, wind turbines can power any number of electrical demands including lighting, charging of electrical appliances, and powering office equipment, to name a few. There are other less obvious uses such as providing electricity for cellular communication equipment or for water pumps in remote locations, for example. In addition to electrical generation, the mechanical energy of a small-scale wind turbine can be used directly to lift, turn, or move objects. While direct use of mechanical energy avoids inevitable losses in the electrical system and can be quite efficient, it is not discussed in much detail here because of the focus on electrical energy generation.

1.4 ADVANTAGES OF SMALL-SCALE WIND OVER COMPETING ENERGY SOURCES

Wind power in general, and small-scale wind power in particular, has a number of features that make it an attractive option for individuals, companies, and other institutions that require power. Perhaps, the most obvious feature is that wind-power systems provide electricity and mechanical energy without reliance on grid-level power systems. Consequently, they provide a level of availability that is often not achieved with grid power, particularly in the developing world.

Second, wind power can be made available at locations that cannot be connected to the grid. These locations, which are often remote, must generate their own electricity and wind should be considered a competitive option. Third, wind power is particularly suited for complementary applications with other renewable energy sources such as solar power to provide a steady source of power as weather and available sunlight fluctuate. Figure 1.7 shows such a combination of wind–solar systems.

Fourth, wind-power systems can be made to be robust. The moving mechanical pieces, if properly balanced and designed, can withstand environmental abuse and mechanical loads for years without degradation. Finally, since the threat of climate change has gained worldwide attention, it has become increasingly clear that human emissions of greenhouse gases, particularly carbon dioxide, are warming the planet [11–13]. Wind power offers a clean source of electricity. Aside from the costs and emissions associated with the construction and erection of a wind turbine, there are no further pollutant emissions. In fact, for some locations around the world, wind-generated electricity can be purchased as an emission offset or the wind energy carbon credits can be sold as an additional revenue stream.

1.5 ISSUES TO CONSIDER PRIOR TO INSTALLING WIND POWER

While the advantages of wind power are clear, there are some issues that must be considered in order to determine whether wind power is viable as an energy-generation source. In particular, the cost comparison of wind

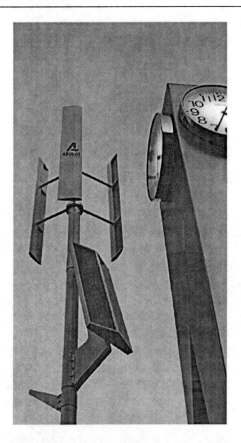

Figure 1.7. Photograph of a Darrieus VAWT connected to a solar panel, positioned in a constrained urban location.

Source: Courtesy of Aeolos Wind Energy, Ltd. (http://www.windturbinestar.com/).

energy against other energy sources is a primary determining factor. Additionally, the siting of wind power in a particular location can have significant impact on the performance of the system, and the subsequent return on investment. While the economic considerations of wind-power installations are a major item addressed later in this book (Chapter 2), some general and cursory statements can be made here.

First, with respect to cost, perhaps, the most simple metric used to assess investment viability is comparison of the cost of electricity from the wind-power system with the costs of electricity from other sources. With this simplistic definition, a cost ratio is obtained.

$$\text{Cost ratio} = \frac{\text{cost of alternative power}}{\text{cost of wind power}} \qquad (1.1)$$

Words of caution are advised here. With wind-power systems, the cost is borne during the design and installation phases, which occur prior to power generation. After installation, the system continues to produce power with little continuing investment. In contrast, energy purchased from other sources have costs that may vary with time, but usually within a small range. Consequently, the cost ratio for wind power decreases with each year of operation; so the expected lifetime must be included in the analysis. Second, when small-scale wind power is used, it is possible to avoid the high costs of power loss that might occur during grid-system failures.

A third item to be considered is that the cost of wind power is heavily impacted by the available wind resources. The energy density of wind varies with the cube of the wind velocity so that a doubling of wind speed leads to an eightfold increase in the flow rate of wind energy. With this recognized, it is seen that proper siting of turbines is crucial for their success. The next chapter will present a more thorough and sophisticated set of performance metrics for wind systems.

Fortunately, wind power maps are available that can guide the siting. For instance, Figures 1.8 and 1.9 show continental wind speeds across the United States at heights of 30 m and 80 m, respectively. A comparison of the figures shows that there is a significant difference in average wind speed for the two heights. This difference often justifies the expense of high-tower construction. Another feature is that the wind speeds are notably higher in the central plains of the United States than on the east or west coasts or downstream from mountain ranges that traverse the western edge of the continent. The central plains are mostly devoid of major blockages to wind flow.

A similar image showing annually averaged wind speeds exclusively for offshore regions is provided in Figure 1.10. It is seen that over water regions, wind speeds are quite high. This generalization also holds for major inland lakes. This behavior is due, in part, to the lack of vertical structures that slow wind near the surface of the Earth. Despite the benefit of higher wind speeds, the installation of a turbine offshore is more complex and expensive than on-land sites.

Similar maps are available for other regions of the world. For instance, Figure 1.11 shows typical wind speeds in China. Generic behaviors that were exhibited in the United States maps (Figures 1.8 and 1.9) are seen to prevail worldwide. Generally, air moves slower over land than over water, particularly for land regions that have vertical structures (mountains, forests, etc.). There are also global patterns of wind speed that are associated with the general circulation of the atmosphere (Hadley cells, for instance) that lead to more appealing conditions in some regions.

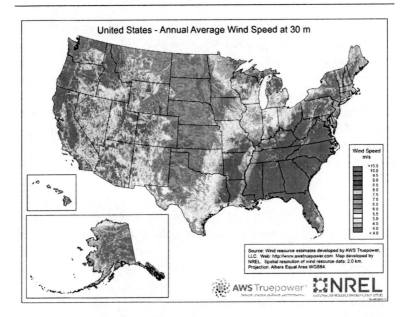

Figure 1.8. Annual average wind speeds at 30 m height across the United States.
Source: Image courtesy of NREL.

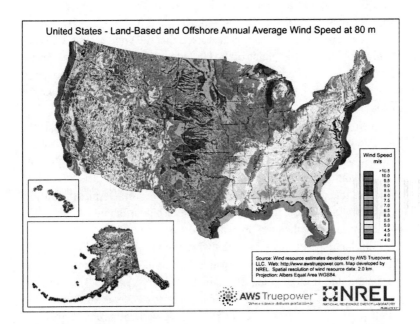

Figure 1.9. Annual average wind speeds at 80 m height across the United States.
Source: Image courtesy of NREL.

INTRODUCTION TO SMALL-SCALE WIND POWER • 13

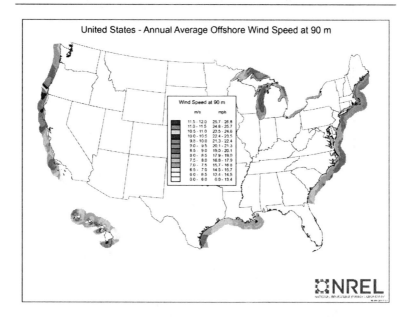

Figure 1.10. Annual average offshore wind speeds in regions surrounding the United States.

Source: Image courtesy of NREL.

Not only are average wind speeds important but so too is the variability of wind. Turbines perform optimally at steady wind speeds that persist for extended durations. Rapid fluctuations of wind, or large variation of daily, weekly, or monthly wind will lower the overall efficiency of the system and in some cases, require control electronics to allow the turbine to adapt to changes in the environment.

There are many sources of wind speed information such as Truepower LLC, Global Atlas, NREL, among others. Attainment of wind information is essential for successful siting and operation of wind-power systems.

As indicated earlier, this book is intended to serve as a reference to persons who are working in the design and development of small-scale wind turbines, are assessing the potential of small-wind systems to provide electricity for a particular application, are evaluating the economic issues related to small-scale wind power, are servicing or installing wind-power systems, or are hoping to forecast the potential of this technology in general. This book is written for both general and specialized audiences, from the homeowner, to the research and development scientist.

Chapter 2 focuses on the economics of small-scale wind systems with a detailed discussion of the issues that should be considered prior to the purchase and installation of a system.

14 • SMALL-SCALE WIND POWER

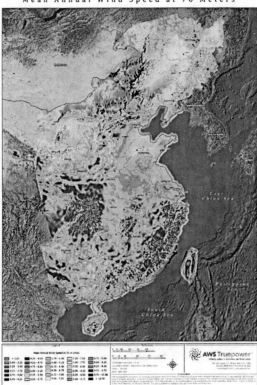

Figure 1.11. Average wind speeds across China, 70 m height.
Source: Image courtesy of AWS Truepower LLC.

Next, in Chapter 3, an exposition on the design of Darrieus-style turbines is given with a summary of existing designs and a discussion of the design, testing, and performance of modern Darrieus rotors.

In Chapter 4, a parallel description of the current status of Savonius-style turbines is given, along with the design and testing procedures and performance measures.

Chapter 5 turns attention to the small-scale HAWTs, which are small versions of the large grid-based wind turbines that are installed around the world.

Chapter 6 describes numerical simulations of small-scale wind turbines, with a special emphasis on the HAWT designs. The simulation process uses advanced computational tools, commonly classified as *computational fluid dynamics* (*cfd*). The justification for the focus on HAWT designs (instead of VAWT designs) is that the latter have been extensively covered in the recent literature, which is listed at the end of Chapter 6.

The final chapter, Chapter 7, brings the discussion together in the form of cases studies. Various examples of wind turbine installations have been taken from around the world to assess the performance of the turbine against expectations. The lessons that were learned from those installations are provided.

REFERENCES

[1]. Abraham, J.P., Plourde, B.D., Mowry, G.S., & Minkowycz, W.J. (2012). Summary of Savonius wind turbine development and future applications for small-scale power generation, *Journal of Renewable and Sustainable Energy*, 4, paper no. 042703.

[2]. Bumby, J.R., & Martin, R. (2005). Axial-flux permanent magnet air-cored generator for small-scale wind turbines, *Electric Power Applications*, 152(5), 1065–1075.

[3]. Higuchi, Y., Yamamura, N., Ishida, M., & Hori, T. (2000). An improvement of performance for small-scaled wind power generating system with permanent magnet type synchronous generator, *26th Annual Conference of the IEEE*, 1037-1043, October 22–28, 2000, Nagoya.

[4]. Tanaka, T., Toumiya, T., & Suzuki, T. (1997). Output control by hill-climbing method for a small-scale wind power generating system, *Renewable Energy*, 12(4), 387–400.

[5]. Whaley, D.M., Soong, W.L., & Ertugrul, N. (2005). Investigation of switched-mode rectifier for control of small-scale wind turbines, *Industry Applications Conference, 40th Annual IAS Meeting*, 4, 2849–2856, October 2–6, 2005.

[6]. Plourde, B.D., Abraham, J.P., Mowry, G.S., & Minkowycz, W.J. (2011). Use of small-scale wind energy to power cellular communication equipment, *Sensors and Transducers*, 13, 53–61, http://www.sensorsportal.com/HTML/DIGEST/P_SI_167.htm.

[7]. Abraham, J.P., Plourde, B.D., Mowry, G.S., & Minkowycz, W.J. (2011). Numerical simulation of fluid flow around a vertical-axis turbine, *Journal of Renewable and Sustainable Energy*, 3, paper no. 033109.

[8]. Dabiri, J.O. (2011). Potential order-of-magnitude enhancement of wind farm power density via counter-rotating vertical-axis wind turbine arrays, *Journal of Renewable Sustainable Energy*, 3: paper no. 043104.

[9]. Kinzel, M., Mulligan, Q., & Dabiri, J.O. (2012). Energy exchange in an array of vertical-axis wind turbines, *Journal of Turbulence*, 13(38), 1–13.

[10]. Mowry, G.S., Erickson, R.A., & Abraham, J.P. (2009). Computational model of a novel two-cup horizontal wind-turbine system, *Open Mechanical Engineering Journal*, 3, 26–34.

[11]. Abraham, J.P. et al. (2013). A review of global ocean temperature observations: implications for ocean heat content estimates and climate change, *Reviews of Geophysics*, 51, 450–483.
[12]. Trenberth, K.E., & Fasullo, J.T. (2010). Tracking Earth's energy, *Science*, 328, 316–317.
[13]. Trenberth, K.E., Fasullo, J.T., & Kiehl. J. (2009). Earth's global energy budget, *Bulletin of the American Meteorological Society*, 90, 311–323.

CHAPTER 2

FINANCIAL AND IMPLEMENTATION CONSIDERATIONS OF SMALL-SCALE WIND TURBINES

B.D. Plourde, E.D. Taylor, P.O. Okaka, and J.P. Abraham

This chapter presents a detailed analysis of the financial considerations that must be given to small-scale wind turbine systems. First, an introduction to the economic issues that govern the viability of wind systems is given. Performance characteristics of the turbines that govern profitability are elucidated and engineering performance metrics are shown. A description of the major costs of wind turbines is provided with a particular focus on those items that are most critical for small-scale systems.

Another major portion of the chapter deals with financial metrics that are used to measure and predict financial performance. Included here are opportunity costs, governmental policies, and energy rebates. Cash flow models are presented showing examples of payback periods (PBPs), return on investment (ROI), net present value (NPV), internal rate of return (IRR), and total cost of ownership (TCO). A case study of a small-scale turbine in Kenya is given.

2.1 INTRODUCTION

Across nearly every region of the globe, small-scale wind turbines have been used in microgrid applications. Microgrids are stand alone power-generation systems that work either autonomously or in conjunction with a utility power grid. The use of local power generation in a microgrid offers a means of providing power in locations where utility-grid power is not financially or technologically feasible. Microgrids create additional advantages over utility grids including reliability, scalability, minimal infrastructure requirements, and they are able to provide power generation from multiple sources. These benefits have stimulated an increased demand for microgrids worldwide. Notably, microgrids have seen tremendous growth in a range of applications including military operations, remote off-grid settings, community utility systems, and industrial markets such as telecommunication base transceiver stations. This chapter covers the fundamental questions that need to be addressed when considering the choice of a small-scale wind turbine for a specific application. The scope of this chapter covers financial and technological considerations related to installing a small-scale wind turbine in a microgrid setting.

Small-scale and microscale wind turbines are defined to have a power production on the order of a few kilowatts during normal wind speeds at the installation location. Of course, the amount of power produced by a turbine is strongly dependent upon the instantaneous wind speed. When wind speeds are very low, little, if any, power is produced. On the other hand, during high wind speeds, large amounts of power are generated. The relationship to power and wind speed will be discussed in greater depth later in this chapter and in the following chapters of the book.

These are wind turbines that are used in situations where traditional, large, horizontal-axis wind turbines (HAWTs) are not appropriate due to the financial and size requirements. The swept area of small-scale wind turbines ranges from 1 to 10 m^2 and the wind turbine systems weigh less than 1,000 kg.

Small-scale wind turbines in microgrid applications are often integrated into a system comprised of complementary power production units, such as solar and diesel. Microgrid structures are able to be scaled and customized to meet specific system requirements. Additionally, the customization of the microgrid allows for each power source to be scaled according to the availability of the local renewable resources. An example of a customized vertical-axis Savonius rotor is shown in Figure 2.1.

Multiple power production components can be integrated into microgrid systems to utilize multiple power generation resources. A complementary relationship between different power generation resources can be seen with wind and solar energy; as the amount of available wind and solar energy changes throughout the day and throughout the year, the power supplied to the microgrid changes. A particular example of this is the northwest hemisphere, where the availability of solar energy is greatest during the summer, and the availably of wind would be greatest in spring, fall, and winter.

Figure 2.1. A computer rendering of a small-scale Savonius VAWT rotor.

With wind and solar power being intermittent, energy-storage systems are commonly found in microgrids. The range of energy storage for a microgrid system is typically from a one- to five-day supply. Microgrids have also been developed for applications where there is reliable utility grid power. These systems are used to reduce the use of utility grid power, for emergency power, or for social, environmental, or marketing reasons.

2.2 SELECTION OF THE WIND TURBINE SYSTEM

There are many different types of small-scale wind turbines that can be used in a microgrid setting, many of which are discussed in this book. Within each different class of wind turbine, there are various orientations and designs of the mechanical supporting structure. These variations came about through the development of small-scale wind across different market sectors. This fact is important to note because different markets require different designs, which have led to a series of different available wind systems. The versatility of different wind turbines has allowed for microgrid designs to be flexible and optimized for specific applications. However, the selection of the correct wind system for a project is critical for the success of the microgrid system. In this section, the differentiation between classes of small-scale wind turbines is identified. With regard to the mechanical components of a wind system, the conceptual design process is also identified.

2.2.1 WIND TURBINE TYPES

As described in Chapter 1, there are two broad classifications of wind turbines, both based on the axis of rotation. Most traditional wind turbines fall into the two classifications: HAWTs and vertical-axis wind turbines (VAWTs). HAWTs are commonly used as megawatt-sized utility grid turbines and have proven to be the most efficient means for converting wind into electrical power. Some HAWTs are designed for small systems, such as the one shown in Figure 2.2. VAWTs are more prominent in small-scale wind applications. Their market advantage comes from being independent of wind direction. This feature has allowed wind-power integration with preexisting structures in areas where physical size is a limitation. For modern VAWTs, there has been a large advancement in the development of drag-induced (Savonius)-style rotors. These designs have shown additional benefits of being able to provide power at lower average wind speed, and with lower rotational speeds.

FINANCIAL AND IMPLEMENTATION CONSIDERATIONS • 21

Figure 2.2. Image of a small-scale HAWT.
Source: Courtesy of Bergey Wind Co.

Darrieus rotors appear as airfoils and are rotated by lift forces. This feature allows a Darrius-style wind turbine to behave as a fast-rotating low-torque machine. Savonius rotors appear usually as concave plates, which are used to capture wind that creates drag forces. The two different mechanisms to produce rotation create a tremendous difference in the system's ability to produce power.

The mechanical and power requirements determine the type of wind turbine to use. Figure 2.3 showcases an example where a series of 5 kW wind turbines are used on an existing structure. The VAWTs shown in the figure allow for the wind system to be independent of wind direction, have a simple installation process, and act as an attractive symbol of green energy. Additionally, the force loading of a Darrieus-style turbine is lower than that of its Savonius counterpart with the same swept area.

Figure 2.3. Small-scale VAWT on Lincoln Financial Field.
Source: Courtesy of Urban Green Energy.

2.3 FINANCIAL CONSIDERATIONS

2.3.1 INTRODUCTION OF ECONOMIC EVALUATION

There are a variety of types and styles of microwind systems, each with unique benefits and ideal implementation scenarios. To decide which system is best, and ultimately to make a purchase decision, an operator must consider the financial and economic factors for their particular application. For a specific microwind system to be chosen, it must be financially superior to other wind systems and power-generation means, or it must provide indirect benefits that justify the additional cost. The main economic factors for microwind systems are commonly set up in a financial model and input into project finance calculations for consistent evaluation.

Among the most important of the financial variables to consider for microwind systems are the initial capital and setup costs. These include the cost of any assets as well as shipping and installation costs that must be incurred for a system to be functional. These costs are typically incurred at the inception of a project and are then amortized over the life of the project. There are some costs of wind systems that are not incurred at the onset of a project and are recurring, such as maintenance, which should also be taken into consideration. Each style, design, and model of a wind system will have its own unique cost characteristics, and therefore, the specific variables associated with each application should be considered.

While costs play an important role in determining the overall financial characteristics of a microwind system, the financial benefit mainly comes from the ability of the system to generate positive cash flow, or reduce negative cash flow, and ultimately generate returns for the operator. Each application will have a unique set of possible sources of return for the operator that should be considered. Perhaps, the most common and directly apparent benefit comes from a reduction in energy costs resulting from energy being drawn from the wind system instead of more costly sources. Other potential sources of return can include revenue generated from selling excess energy back to the grid through a utility buyback arrangement, reduction in tax liability for the operator through energy credits and other tax credits, government subsidies and rebate programs, carbon credits on the open market, as well as decreases in expense due to reduction in system downtime.

2.3.2 PERFORMANCE CHARACTERISTICS THAT IMPACT RETURNS

While there are many considerations when evaluating the use of a wind system, the most fundamental variables are the prevailing wind speed at the specific site and the performance of the wind turbine at different wind speeds. While these variables do not make up the full economic impact, they are often the first variables considered when evaluating a potential application.

2.3.2.1 Prevailing Wind Speed

To fully evaluate the economic impact of a wind system the most basic variable to consider is the prevailing wind speed at the proposed installation site. Wind speed data for various geographic locations worldwide are published by numerous educational, nonprofit, and for-profit organizations. When assessing wind speed, however, complete reliance cannot always be given to published wind data, as there are several considerations that impact the actual characteristics of the wind.

Elevation is an important consideration as prevailing wind speed generally increases with height. Wind data published for a particular location can often only serve as a proxy if the particular application calls for installations above or below the elevation of the wind data. There are also physical objects that can both positively or negatively impact wind speeds. Some wind system designs and installations have attempted to strategically take advantage of objects by installing systems such that they are in areas where winds speeds are higher than the undisturbed surroundings. Wind speeds also change throughout the year, and they can change for

24 • SMALL-SCALE WIND POWER

the same location and same season year-over-year; so it is best to look at historical averages and patterns, as well as the expected environmental shifts over time to fully assess and determine the expected wind speeds for a particular location.

Additionally, published wind speeds are averages, and the wind often varies significantly throughout the day and year. Since the power curve is nonlinear, and more power is generated at higher wind speeds, it is important to consider how often the wind is not blowing, and how often the wind is blowing at above-average speeds. If a wind system is capable of capturing wind gusts that are higher than the average published wind speed, the overall power produced will be higher than it would be if the wind blew constantly at the average wind speed. Since wind speed variances are rarely published, this factor is often ignored when calculating the actual power production expected or the true financial return of a wind system.

In order to gauge the actual wind characteristics expected at a particular installation location, it is common practice to install an anemometer to make direct measurements. These measurements can then be compared to published wind data for the area. For many areas, there are strong seasonal changes in prevailing winds; in these circumstances, measurements must be made over sufficiently long times.

2.3.2.2 Power Curve

Another essential factor for predicting the performance of a wind system is the power production and wind speed relationship, commonly known as

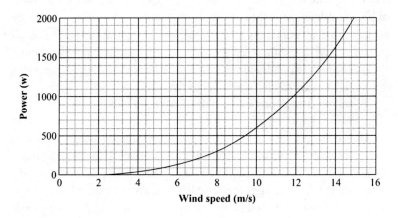

Figure 2.4. Power curve for small-scale wind turbine with a swept area of 5 m² and a C_p of 0.2.

the power curve. This measure is appropriate as the relationship is nonlinear and the power production increases with respect to wind speed at an increasing rate. A typical power curve is shown in Figure 2.4. The graph has been created with a coefficient of performance $(C_p) = 0.2$. The coefficient of performance is a measure of the production of energy compared to the available energy flowing in the wind.

Specific power curves are unique to each model of wind turbine, and as discussed earlier in this chapter, different wind turbine types have different power curves. The manufacturer or distributor of the wind turbine typically conducts extensive tests, both in-the-field and in wind tunnels, and publishes this curve. It is important to note that there is a significant difference between the maximum rated power, which is the maximum amount of power the generator on the wind system can produce, and the power curve.

2.3.3 COSTS OF SMALL-SCALE WIND TURBINES

Small-scale wind turbines can provide great financial benefit in a variety of scenarios. One part of quantifying the overall return of a system over time is through its cost characteristics. As each style and model of wind turbine has different cost levels and considerations, it is important to understand the potential costs associated with a wind turbine: system costs, shipping, assembly and installation, and maintenance costs.

2.3.3.1 System Cost:

The cost of a wind system is commonly defined as the purchase price of raw materials and preassembled components used for the system. This is a one-time initial investment, which includes wind turbine rotors, mechanical structures, mast, and attachments between the mechanical structure and the mast.

The cost of a wind system should not be considered in isolation, however. Wind turbines made with advanced materials typically have a higher cost, but also typically have favorable mechanical characteristics that can lead to longer useful lives and less required maintenance. Lower-cost systems are often made with lower-quality components and may be more expensive to operate over the life of the system due to increased maintenance and downtime due to failures. Some wind turbines include warrantees against failed systems as a part of their price, which increases the initial cost of the system, but can reduce or eliminate potential future maintenance and replacement costs.

2.3.3.2 Shipping

Shipping is a cost category that is often underestimated. By nature, microgrid wind applications are often considered in areas where there is no well-developed transportation infrastructure, and therefore, it is important to consider the cost of shipping all the way to the final installation site. Shipping costs correlate highly with package size and weight. Consequentially, some systems are designed such that several units can fit into standard-sized shipping crates, which make deploying several systems in closer geographic proximity more economical.

2.3.3.3 Assembly and Installation

Assembly and installation consist of setting up the wind system from a packaged state in its shipping container to its final, fully functioning, power-producing state. These costs predominately consist of labor and machinery costs. Assembly and installation are typically considered together, as it is most economical and cost effective for assembly and installation to both be done at the installation site, on the same day, and by the same crew.

The equipment required to install a wind turbine can significantly impact the installation expense. The cost of using a crane, for example, is several orders of magnitude greater than that of a truck winch. Similarly, labor costs are a function of skill and time, and therefore, a simple and quick assembly and installation process is desired. More complex assembly and installation processes take longer and require higher skilled labor, which is more costly.

Another consideration of labor is whether to use in-house labor or local labor. The most economical choice is often local labor for the installation, but this depends on the location of installation and the distance that must be traveled by the crew. Regardless of cost, it is important to consider the possibility that out-of-house labor may not be as familiar with the assembly and installation process as in-house labor, and therefore is less reliable.

Figure 2.5 shows a wind system designed for a telecommunication base transceiver station. This is an example of a system designed for a specific application. The installation process of this unit was designed to mimic the installation process of other components commonly installed on telecommunication towers. This feature was intentionally designed to avoid the costs associated with setting up an independent tower, and to take advantage of the availability of crews familiar with the process and equipment required to install telecommunication equipment on towers.

Figure 2.5. A small-scale wind turbine designed for telecommunication base transceiver stations during installation on a USA Department of Transportation telecommunication tower, 2012.

2.3.3.4 Maintenance

Maintenance is the only cost category that is ongoing throughout the life of the system. Maintenance is the cost of labor, tooling, materials, maintenance checks, preventative replacements or repairs, parts, and other ongoing costs incurred to keep the unit in working order. Maintenance, unlike the other points discussed in this section, has a direct and ongoing negative impact on the economics of a system, and does not get amortized over time as the other capital costs do.

Different styles and models of wind systems have different maintenance requirements, and maintenance schedules can vary widely between systems. Multiple factors such as components, materials, and system design all impact maintenance requirements. Manufacturers and distributors typically publish good faith estimates on the amounts of maintenance their systems require in numbers of hours of service required per year. Although this information may be provided, it is important to consider the external factors, such as weather, which can have an impact on how long systems can perform before maintenance is required. For most microgrid wind systems, bi-yearly or yearly maintenance should be forecasted.

2.3.4 FINANCIAL RETURNS OF SMALL-SCALE WIND TURBINES

Although costs are very important to consider when evaluating a microgrid wind system, they only make up part of the total economic impact. There are several different sources of financial benefit from microwind systems that ultimately drive financial returns. These returns can come from decreased spending on the source of power that is being replaced, arrangements with the local utility or government whereby they purchase surplus power from a system, subsidies from local or national governments, energy credits and other favorable tax implications, carbon credits that can be sold to third parties, as well as government rebates to help offset the cost of equipment and setup.

2.3.4.1 Reduction of Energy Expenditure (Opportunity Cost of Energy)

The main goal of any microgrid wind system is to provide power that is generated from a renewable source with minimal variable costs. Where wind systems are deployed to displace or reduce the reliance on an existing power source, the most desired return is from the reduction of energy expenditure compared to the previously used power source.

If the wind system is replacing a current source of power, the total current cost of power is used to determine the reduction in energy expenditure. If, for example, a generator is the current source of power, the current cost of power would include the amortized capital cost of the generator over its life time power production, the cost of fuel that is used in the generator to produce each unit of power, the transportation costs to get fuel to the generator, and any periodic maintenance required to keep the generator in working order. All costs associated with the existing source of power that are reduced or eliminated by a wind system should be considered in the return of that system. For example, if the cost of fuel lost to theft or the security to prevent theft is a cost that is occasionally incurred in some locations, it should be included in the analysis. Similarly, any security costs associated with the wind-turbine system also must be considered.

For an example of the reduction in energy expenditure, assume that the total cost of energy for a system being replaced is X USD/kWh. For every kWh that is produced by the wind system instead of this source of energy, X USD is a direct return from the wind turbine system. Table 2.1 provides a numerical example of the return from the reduction in energy expenditure.

In Table 2.1 it is assumed that the wind system generates 0.5 kW continuously. This means that over a 24-hour day, the system will generate 12 kWh of energy. If the total cost of energy being replaced is $0.45/kWh, the daily savings the wind turbine provides is $5.40. As can be seen from the last lines in this example, these savings can be quite significant when considered over a longer time period.

An important consideration when evaluating a microwind installation is the maximum potential benefit from reducing energy expenditure, which is limited by the power requirements and current cost of power for the specific application. The economic benefit from reducing the current energy expenditure can only be as high as the total current cost of energy. Unless a utility buyback arrangement is available, any time the wind system is producing more energy than the application is using, there is no extra benefit.

Therefore, the benefit from using a wind turbine instead of an existing energy source cannot be greater than the elimination of 100% of the current total power cost of the application. This means that the size of the microgrid system must fit with the application, as a more expensive system that is likely to generate excessive power will likely not have financial characteristics that are as attractive as one appropriately sized for the application. This consideration is mainly applicable when the specific application is

Table 2.1. Numerical example of the return of a microwind turbine resulting from a reduction in energy expenditure

Performance Measure	Quantity	Units
Average power produced	0.5	kW
Average energy produced daily	12	kW/day
Average previous cost of power	$0.45	USD/kWh
Average daily savings	$5.40	USD
Average monthly saving	$162	USD
Average annual saving	$1,971	USD

intended to serve as a replacement or supplement to the existing power. If the wind system is being installed where there is currently no power, as a backup system in case of power outages, or in a situation where a utility buyback can be arranged, then this consideration is not required.

2.3.4.2 Rebates

The initial investment of a wind system typically requires a large capital outflow that is amortized over the life of the system; however, there are occasionally rebates available from governments or other organizations to reduce the initial cash flow burden and make wind systems more financially attractive. The purpose of these programs is to incentivize wind system deployments for their long-term economic, infrastructure, and environmental benefits for the area. Rebate availability and details vary widely by sponsoring organization and installation locations. Rebates may not be available in some areas; they may be as small as eliminating sales tax on system purchases, or in some cases, the rebate amount may account for a majority of the capital cost of the wind system. It is advisable to check the availability of rebates from governmental programs and other organizations for each application being considered. Since rebate programs do not have indefinite timelines, it is prudent not to rely on them for installations far in the future.

2.3.4.3 Utility Buyback Arrangements

When wind systems are installed in areas or applications where available connection to a utility grid exists, it is possible that a buyback arrangement may be available. These arrangements work when the utility pays for or provides credit for excess energy produced by a microgrid system that is sent back to the grid. In recent years, laws have been passed in some areas that require the utility company to accept and pay for energy surpluses generated from decentralized sources. Equation 2.1 shows a simple formula for the potential cash flow from energy sold back to the grid. Sometimes, utility buyback programs can have more complex pricing structures. It is advisable to check both the availability and specifics for each particular application.

$$\text{Utility Buy back Proceeds} = \text{Energy Surplus Sold} \times \text{Utility Buy back Rate} \qquad (2.1)$$

It must be noted that any expected energy surplus needs to account for variations in wind speed and power generation. It is common to use average values of surplus energy production in this calculation.

2.3.4.4 Government Tax Credits and Subsidies

Although much more common with megawatt-sized wind farms, government subsidies can help enhance the financial return of microwind systems. A wind subsidy is a payment to wind turbine operators that helps to offset the cost of power produced by a wind system. Although there are various reasons subsidies exist, they are usually intended to provide financial support for projects that are determined to have positive social or economic benefits.

Wind tax credits are a sum deducted from the tax burden, or amount owed in taxes. They are typically correlated with the size of the wind system installation or the amount of energy produced. Some government programs allow entities to write off some portion of the capital expenditure for a wind system deployment. Specific details and amounts allocated to these programs vary widely by government, location, and over time. While subsidies and tax credits can greatly enhance the financial characteristics of a wind system in any location, it is important to consider microwind applications without government subsides as their effective duration is often shorter than the intended life of a microwind turbine, and they are subject to the risk of nonrenewal.

2.3.5 CASH FLOW MODEL FOR FINANCIAL EVALUATION

For a full economic evaluation, all of these variables discussed must be considered and analyzed. The most common way this is done is by determining the likely periodic cash flows from each of these variables over the estimated life of the system. An example of this cash flow model can be seen in Table 2.2, where cash flows for the operator are represented by negative symbols (−) representing cash outflows, and positive symbols (+) representing cash inflows. As visualized in this table, the capital and setup costs associated with a wind turbine project are incurred at the onset of the project, with maintenance typically being the only continuing negative source of cash flow throughout the life of the system. The main sources of positive cash flow that are possible will ideally last throughout the life of the system and are what make up the return for the operator.

Table 2.2. Cash flow model example for small-scale wind turbines

Cash Flow Model	Initial	Year 1	Year 2	Year 3	Year (n)
Cost of system	−				
Rebates	+				
Shipping	−				
Assembly	−				
Installation	−				
Maintenance		−	−	−	−
Energy reduction		+	+	+	+
Utility buyback		+	+	+	+
Tax credit		+	+	+	+
Subsidies		+	+	+	
Energy credits		+	+	+	
Carbon credits		+	+	+	
Total of all cash flows					

Each application will likely have a specific set of expected cash flows that result from these different categories; so a unique model should be developed for each case.

2.3.6 FINANCIAL CALCULATIONS FOR PROJECT EVALUATION

While forecasting cash flows of a particular wind system is the first step for cash flow analysis, it is important to enter these forecasted cash flows into common project finance calculations and consider the guidance they provide before making a decision. The most common calculations that are used when evaluating a microwind project are the PBP, the ROI, the NPV, the IRR, and the TCO. These calculations and the guidance they provide are widely used to determine investment decisions for a variety of capital investments.

2.3.6.1 PBP

One interpretation of the PBP is the length of time it takes for a system to generate enough positive cash flow, or reduce enough negative cash flow, to recover the initial capital costs of setting up a system. The PBP is one of the most common calculations used to evaluate the potential attractiveness of a microwind system. In practice, it is often the first variable calculated and considered, and some wind turbine vendors will provide PBP estimates for their systems. The PBP calculation is shown in Equation 2.2.

$$\text{Payback period} = \frac{\text{Capital Expenditure}}{\text{Net Change In Periodic Cash Flow}} \quad (2.2)$$

Acceptable ranges for the PBP vary widely among different geographic locations and for different applications. A PBP under 5 y is ideal in less-developed areas where the price of electricity is higher (typically over $0.30/kWh) or where the utility grid is not well established or reliable. In more developed countries where there are reliable grids and lower energy prices, the PBP threshold can be 8 y or longer. Typically, systems are not considered when the PBP is greater than 10 y. The market dynamics that determine energy prices change frequently, and the available technology to deliver energy is too variable to justify the investments that will not be recovered until that far in the future.

2.3.6.2 ROI

The ROI calculation is a financial evaluation metric used to gauge whether the return on a particular investment compares favorably to the investment cost. The equation to determine ROI from cash flows over the project life is shown in Equation 2.3.

$$\text{ROI} = \frac{\text{Gain(s) From Investment} - \text{Cost of Investment}}{\text{Cost of Investment}} \quad (2.3)$$

ROI is used most often to evaluate the overall efficiency of a proposed project or investment, or compare the overall efficiency of several investments in consideration. Economically, it considers profits in relation to capital required for an investment, and therefore projects with a higher ROI should be held in higher favor than projects with a lower ROI, and projects with a negative ROI should not be considered.

The main drawback to ROI is that it does not account for how far in the future cash flows are expected. Using the same life cycle time for all the options being compared will eliminate this problem.

2.3.6.3 NPV

The NPV is the sum of the present value of cash inflows and outflows over the project life. It is a financial metric used to analyze and evaluate the profitability of an investment or project, with specific sensitivity to cash flows forecasted further in the future, which are considered to be less certain. The NPV is calculated using Equation 2.4.

$$\text{NPV} = \sum_{t=0}^{T} \frac{C_t}{(1+r)^t} \qquad (2.4)$$

In Equation 2.4, C_t is the net cash flow in a specific period; t is the time period in years, from the initial period of t equal to zero, through the final period represented by T; and r is the discount rate, which is commonly set at 10% for initial calculations. Systems that have a positive NPV will add economic value and should be considered. Additionally, when considering two different systems, the one with the higher NPV should be considered superior as it is expected to add more economic value.

2.3.6.4 IRR

The IRR is a financial project evaluation measure that helps to determine the relative profitability of proposed projects. It works similar to the NPV; however, instead of manually determining the discount rate, as is done when finding NPV, the IRR is the specific discount rate that makes NPV equal to zero. It is calculated in Equation 2.5.

$$\text{NPV} = \sum_{t=0}^{T} \frac{C_t}{(1+IRR)^t} = 0 \qquad (2.5)$$

In Equation 2.5, C_t is net cash flow in a specific period; t is the time period in years, from the initial period of t equal to zero, through the final period represented by T; and IRR is the discount rate that makes the NPV equal to zero. When other factors are equal, it is advisable to select the project with the higher IRR, as this means that the project's expected future profitability is greater. However, it is important to consider that since the structure of this formula assumes that all received cash flows are reinvested at the IRR rate, results are not linear with IRR, and therefore an IRR for one project that is twice as high as another does not mean that the project is twice as profitable. A word of caution is given that there is potentially serious negative impact of uncertainty in the estimation of future values.

2.3.6.5 TCO

The TCO is a financial evaluation tool that helps evaluate the overall costs, from both direct and indirect sources, of a proposed project or of multiple proposed projects. While TCO does not give a clear distinction of a project being more economically favorable, or whether a project should be accepted or not, it is an effective way to analyze the full scope of costs when considering and deciding on courses of action for a project. The calculation of TCO is shown in Equation 2.6.

$$\text{TCO} = \sum_{t=0}^{T} \text{Purchase Price} + \text{Direct Costs} + \text{Indirect Costs} \qquad (2.6)$$

The TCO calculation is used only when comparing the total cost of proposed projects for the entirety of the project life. It is mainly considered in conjunction with the total budget for a particular application.

With the major turbine variants described, the financial metrics provided, and performance discussed, it is possible to showcase an actual planned installation of a small turbine. The lessons learned in the preceding section are employed.

2.4 CASE STUDY—TELECOMMUNICATION TRANSCEIVER TOWERS IN KENYA, AFRICA

The case study in this section shows the evaluation process of a proposed microwind system in Kenya, Africa. The considerations and methods for evaluating a wind turbine are evaluated and discussed in this case. Actual data and constraints of the application and industry are used.

2.4.1 Background

Telecommunication equipment operators in Africa provide and maintain telecom towers for wireless providers throughout the continent. As market saturation has increased in urban areas, telecom companies have attempted to provide telecommunication service in a broader set of locations to grow market share. Since the development rate of telecommunication infrastructure in rural areas significantly outpaces development of government utility grid infrastructure, telecom equipment has increasingly been deployed in locations where power grids are not established or are not reliable. Currently, the most widespread solution to this problem is to provide power though the use of diesel generators. Diesel generators are not ideal as they produce toxic exhaust and greenhouse gas emissions, and often have higher per unit energy costs compared with grid prices.

2.4.1 PERFORMANCE CHARACTERISTICS

The first step of the evaluation process consisted of examining the proposed wind turbine performance characteristics as well as the wind characteristics at the proposed site.

2.4.1.1 Power Curve

The power curve for the proposed system is shown in Figure 2.6. This power curve is best represented by having a C_p of 0.3 and the swept area of the turbine is 6 m² (63 ft²). This power curve is considered above average for Savonius-style wind systems with similar swept area.

2.4.1.2 Prevailing Wind Speed

Published data on wind speeds for this particular location are available through Global Atlas (available at www.irena.org), which is an open source database for wind maps and other renewable energy resources. A wind map of the area from this source is shown in Figure 2.7. The proposed installation height was consistent with the height of the published wind data. The deployment was in a remote location to serve as a pilot project, and an anemometer was not set up in advance to gather further data on the wind speed; so complete reliance was shown toward published wind data. Based on this wind speed information, average wind speeds of 8 m/s (18 mph) were expected.

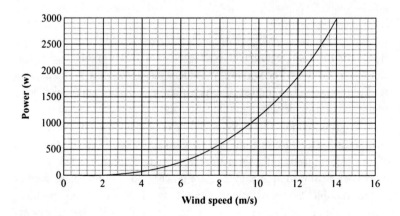

Figure 2.6. Power curve for small-scale wind turbine with a swept area of 6 m² (63 ft²) and a C_p of 0.3.

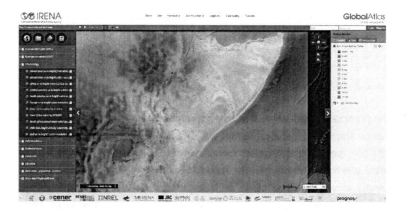

Figure 2.7. Average wind speed map of Kenya.
Source: Courtesy of Global Atlas.

2.4.1.3 Performance Conclusions

From the power curve and wind speed considerations previously discussed, it was assumed that the average power produced would be 600 W. This figure was in line with the expectations of the operators considering the deployment; so cost considerations were evaluated next.

2.4.2 COST CONSIDERATIONS OF THE APPLICATION

The next consideration evaluated was the cost characteristics of the proposed system.

2.4.2.1 System Cost

The manufacturer of the proposed wind system sold complete units as a single packaged item. Their 3 kW rated system, with a power curve shown in Figure 2.6, included a two-section rotor, axle, mechanical structure, generator, electronics, and an accelerometer, and was priced at $10,500. The system had a design feature that allowed it to be attached to existing structures. This allowed it to be installed at a significant elevation on an existing structure and eliminated the need for a separate mast to be erected.

Some of the integral components of the wind system were made with advanced materials and manufacturing processes that favorably impacted its durability and expected useful life compared with other wind systems.

Other parts of the system were intentionally "off-the-shelf" to reduce potential replacement costs.

2.4.2.2 Shipping

The proposed application was in a remote region of Africa. This location was far from the US-based manufacturing facility of the vendor. The local area had poor transportation infrastructure surrounding it, which made travel and transportation all the way to the proposed installation site more complicated and less reliable. To transport the unit from the manufacturing facility in the United States to Nairobi, Kenya, which was the closest major city to the proposed installation site, air freight was considered to be the most economical option. For the specific size and weight of the shipped package the shipping quote was $2,557. Although other shipping methods were evaluated, this quote was the most economical, and came with insurance.

To transport the unit from the shipping facility in Nairobi to the installation site, local transportation costs were found to be $140.

2.4.2.3 Assembly and Installation

The proposed wind system was designed to be assembled by two people without the need for power tools. The wind system, as discussed previously, had a unique design in that it was able to attach to existing structures. This enabled it to be installed with a truck winch, avoiding the expense of a crane, which can not only be difficult to obtain in some areas, but also can cost thousands of dollars per installation. For this application, local labor was chosen, as there was an available crew that was familiar with the telecom site. Using two crew members and a light truck, assembly and installation were able to be completed in one day, and cost $478.

2.4.2.4 Maintenance

The wind turbine proposed for this case had maintenance recommendations that consisted of annual inspections and biennial preventative maintenance procedures. The annual inspections consisted of an engineer visually inspecting the mechanical structure as well as the electronic components in the housing, and the biennial preventative maintenance consisted of the attachment points being checked, and the bearings being checked and greased. Each year, the average maintenance forecasted for these checks was $250.

2.4.3 RETURN CONSIDERATIONS OF THE APPLICATION

The proposed site in this case had several characteristics that contributed to the attractiveness of the potential financial return.

2.4.3.1 Reduction of Energy Expenditure

The first step in calculating the potential return from reducing energy expenditure has already been completed. It was determined in the Performance Characteristics section that the average expected power of the system was 600 W. The other main variable that must be calculated is the current cost of power. At the site in this application, a diesel generator was used as the main source of power. From historical cost records of diesel purchases, transportation of fuel to the site, maintenance of the generator, and the original amortized cost of the generator itself, it was determined that the total cost of fuel was $0.55/kWh. No adjustments were needed for other factors such as theft or security, as there were no records of costs being incurred from them. The average energy required at the site was 1 kW; so it was assumed that all of the power produced by the wind turbine would go toward eliminating current energy costs, and that producing more power than could be used or stored was not a risk. With this information, an estimation of the return from the reduction in energy expenditure could be created. As shown in Table 2.3, the annual reduction in energy expenditure was estimated to be $2890.80.

Table 2.3. Numerical example of the return of a microwind turbine resulting from a reduction of energy expenditure

Performance Measure	Quantity	Units
Average energy produced	0.6	kW
Average energy produced daily	14.4	kW/day
Average previous cost of power	$0.55	USD/kWh
Average daily savings	$7.92	USD
Average monthly saving	$237.60	USD
Average annual saving	$2,890.80	USD

2.4.3.2 Rebates, Utility Buyback Arrangements, and Government Tax Credits and Subsidies

The application in this case was in an area where no rebates or government tax credits were available, and since the site was not connected to a utility power grid, a utility buyback arrangement was not possible. Therefore, no cash flows from these sources were expected.

2.4.4 CASH FLOW MODEL FOR FINANCIAL EVALUATION

With the all performance, cost, and return considerations available, a financial model was built to forecast yearly cash flows of the potential system. This cash flow model can be seen in Table 2.4, which is a counterpart to Table 2.2.

2.4.5 FINANCIAL CALCULATIONS FOR PROJECT EVALUATION

With the financial model of all forecasted cash flows built, financial project evaluation calculations could be determined and evaluated.

2.4.5.1 PBP

For this application, the operator established acceptable parameters for the PBP at less than 7 y. Based on the forecasted project cash flows for this application, the PBP was calculated by Equation 2.2 to be 5.18 y.

$$\text{Payback Period} = \frac{\$10,500 + \$2,557 + \$140 + \$478}{\$2,891 + \$250} = 4.35 \text{ y} < 7 \text{ y} \quad (2.7)$$

This PBP met operator expectations of less than 7 y for a payback.

2.4.5.2 ROI

For similar infrastructure capital investments, the operator used a 10 y view when evaluating ROI. Therefore, 10 y of cash flow were considered in the ROI calculation, taken from Equation 2.3 to be

$$\text{ROI} = \frac{(\$2,641 \times 10 \text{ y}) - \$13,680}{\$13,680} = 0.93 \text{ or } 93\% \quad (2.8)$$

Table 2.4. Cash flow model example for a small-scale wind turbine

Cash Flow Model	Initial	Year 1	Year 2	Year 3	Year 4	Year 5	Year 6	Year 7	Year 8	Year 9	Year 10	Sum
Cost of system	10,500											10,500
Less: rebates	–											–
Shipping	2,702											2,702
Assembly and installation	478											478
Ongoing maintenance		250	250	250	250	250	250	250	250	250	250	2,500
Reduction in energy costs		2,891	2,891	2,891	2,891	2,891	2,891	2,891	2,891	2,891	2,891	28,908
Total cash flow	13,680	2,641	2,641	2,641	2,641	2,641	2,641	2,641	2,641	2,641	2,641	12,728

Although no acceptability parameters were established beforehand, an ROI of 93% was favorable to the operator for this application over the 10 y horizon, and therefore, it contributed positively to the installation decision.

2.4.5.3 NPV

To calculate the NPV, the operator had to determine the discount rate, and the number of expected years of operation. A discount rate of 10% was used and the number of periods was capped at 10 y. NPV, calculated over 10 y from Equation 2.4, is shown as follows.

$$NPV = \frac{-13,680}{(1+.1)^7} + \frac{\$2,641}{(1+.1)^1} + \frac{\$2,641}{(1+.1)^2} + \frac{\$2,641}{(1+.1)^3} + \frac{\$2,641}{(1+.1)^4} + \frac{\$2,641}{(1+.1)^5}$$
$$+ \frac{\$2,641}{(1+.1)^6} + \frac{\$2,641}{(1+.1)^7} + \frac{\$2,641}{(1+.1)^8} + \frac{\$2,641}{(1+.1)^9} + \frac{\$2,641}{(1+.1)^{10}} \quad (2.9)$$
$$= \$2,546.57$$

Although details on other infrastructure investments being considered by the operator were not disclosed, the operator considered an NPV of $2,546.57 to be acceptable for this project, and therefore, it contributed positively to the investment decision.

2.4.5.4 IRR

The operator did not disclose the type of other capital investments that were available to him at the time; however, it was determined that an IRR greater than 10% was desired. In addition to the number of periods the operator used when calculating NPV, 10 y was also used as the maximum number of period for the IRR calculation. The IRR is calculated using Equation 2.5 and the result is shown in Equation 2.10.

$$0 = \frac{-13,680}{(1+.1418)^0} + \frac{-\$2,641}{(1+.1418)^1} + \frac{-\$2,641}{(1+.1418)^2} + \frac{-\$2,641}{(1+.1418)^3}$$
$$+ \frac{-\$2,641}{(1+.1418)^4} + \frac{-\$2,641}{(1+.1418)^5} + \frac{-\$2,641}{(1+.1418)^6} + \frac{-\$2,641}{(1+.1418)^7} \quad (2.10)$$
$$+ \frac{-\$2,641}{(1+.1418)^8} + \frac{-\$2,641}{(1+.1418)^9} + \frac{-\$2,641}{(1+.1418)^{10}}$$

The calculated IRR to solve Equation 2.10, determined by iteration, is found to be 0.1418 (14.18%). This is greater than the 10% parameter determined by the operator; therefore, this calculation also contributed positively to the investment decision.

2.4.5.5 TCO

For this application, the operator did not consider TCO as a necessary project finance metric to consider. It was therefore not calculated or evaluated in this application.

2.4.5.6 Decision

After the evaluation of the performance, cost, and return characteristics, the operator can decide to proceed with the present installation.

2.5 CONCLUDING REMARKS

In this chapter, many of the issues that must be considered prior to deployment of wind turbine systems are discussed. This includes both performance and financial metrics. Many of the most common metrics were discussed and then applied to an actual application. It was found that in the case study presented here, the turbine under consideration presented a positive choice for the operator.

REFERENCES

[1]. Abraham, J.P., Plourde, B.D., Mowry, G.S., & Minkowycz, W.J. (2011). Numerical simulation of fluid flow around a vertical-axis turbine, *Journal of Renewable and Sustainable Energy*, 3, paper no. 033109.
[2]. Abraham, J.P., Plourde, B.D., Mowry, G.S., & Minkowycz, W.J. (2012). Summary of Savonius wind turbine development and future applications for small-scale power generation, *Journal of Renewable and Sustainable Energy*, 4, paper no. 042703.
[3]. Plourde, B.D., Abraham, J.P., Mowry, G.S., & Minkowycz, W.J. (2011). Use of small-scale wind energy to power cellular communication equipment, *Sensors and Transducers* 13, 53-61.

[4]. Global Atlas. (2013). *International renewable energy agency.* Retrieved September, 2013, from Global Atlas database.
[5]. Bergey Wind Co. (2013).
[6]. Urban Green Energy Inc. (2013).

CHAPTER 3

DESIGN OF DARRIEUS-TYPE WIND TURBINES

M. Islam

This chapter presents an up-to-date resource for the design and analysis of Darrieus-style wind turbines. These turbines differ from the more traditional horizontal turbines in that they spin about a vertical axis. They can be used in constrained areas where large turbines are not viable. Air flow turns the turbine by using lift forces so that the spin occurs at a high rate and excellent power generation can be achieved. The chapter focuses on small-scale Darrieus turbines that can be used for local energy generation in commercial or residential applications.

The chapter covers all major design aspects for these types of turbines and fully reviews the literature. In addition, a discussion is given of various investigative techniques such as experimentation and numerical simulation. It concludes with a brief discussion of recent modification of Darrieus designs including twisted and hybrid variants.

3.1 INTRODUCTION

As discussed earlier, modern-day vertical-axis wind turbines (VAWTs) are generally classified as Darrieus and Savonius types. This chapter focuses mainly on the design aspects of Darrieus-type VAWTs. The original Darrieus-type VAWTs were invented by George Jeans Mary Darrieus, a French engineer, who submitted a patent in 1931 in the United States [1]. The patent of Darrieus included both "Eggbeater (or Curved-bladed)" and "Straight-bladed" type VAWTs, which are illustrated in Figures 3 and 6 of the patent, respectively. However, according to Templin and Rangi [2], they reinvented both the curved- and straight-bladed Darrieus-type VAWTs in 1966 while working as research officers at the National Research Council of Canada.

Basically, the blades of the Darrieus-type VAWTs, as shown in Figures 3.1 and 3.2, are driven by lift forces. The turbine consists of multiple airfoil-shaped blades that are attached to a rotating central column or vertical shaft. The air flow over the blades of VAWTs generates aerodynamic lift that rotates the blades. Both the horizontal-axis wind turbines (HAWTs) and Darrieus-type VAWTs have their strengths and weaknesses. The two main advantages of Darrieus-type VAWTs over HAWTs are: (i) they can operate with incoming wind velocity from any direction unlike HAWTs, which require a yaw mechanism, and (ii) the generator and gear box of the Darrieus-type VAWTs can be placed on the ground. However, these two and other numerous advantages of VAWTs are overshadowed by

Figure 3.1. Darrieus rotor near Heroldstatt in Germany.

Source: Page URL: http://commons.wikimedia.org/wiki/File%3ADarrieus_rotor002.jpg; File URL: http://upload.wikimedia.org/wikipedia/commons/9/92/Darrieus_rotor002.jpg; Attribution: By W.Wacker (Own work) [CC-BY-SA-3.0 (http://creativecommons.org/licenses/by-sa/3.0)], via Wikimedia Commons.

Figure 3.2. Straight-bladed Darrieus VAWT (SB-VAWT).

two main disadvantages of Darrieus-type VAWTs, which are (i) self-starting problem; and (ii) oscillating loads on the support structures. The interested reader can refer to Refs. [3–7] for comparative analysis between VAWTs and HAWTs. Particularly, Eriksson et al. [4] conducted a detailed comparative study between two different types of Darrieus-type VAWTs (namely eggbeater and straight-bladed) and HAWTs and concluded that "*VAWTs are advantageous to HAWTs in several aspects.*" Islam et al. [6] estimated that VAWTs can dominate the wind-energy technology within the next two to three decades. It should be noted that this chapter reviews different aspects related to the design of smaller-capacity Darrieus-type VAWTs only. One of the principal goals of this chapter is to provide a comprehensive literature review of all aspects of Darrieus-style turbines.

3.2 CURRENT DARRIEUS TURBINE DESIGNS

Comprehensive studies on various VAWT configurations and design techniques are found in Bhutta et al. [5] and MacPhee and Beyene [8]. Bhutta et al. [5], have identified 11 different types of VAWTs that have varying degrees of advantages and disadvantages. The main feature of the eggbeater-type Darrieus VAWT, as shown in Figure 3.1, is that it minimizes the bending stress in the blades. However, the fabrication of blades is more challenging. In contrast, the straight-bladed-type Darrieus VAWT (often called giromill or H-rotor), as shown in Figure 3.2, has very simple geometric features and more widely used for remote and urban applications.

The straight-bladed Darrieus-type VAWTs (SB-VAWTs) fall into two categories: (i) fixed pitch and (ii) variable pitch. Based on previous research works, the fixed pitch SB-VAWTs provide inadequate starting torque [9]. Whereas, the variable pitch blade configuration has potential to overcome the starting torque problem; however, it is quite complicated and impractical for smaller-capacity applications.

3.3 DESIGN OF DARRIEUS-TYPE VERTICAL-AXIS WIND TURBINES

3.3.1 DESIGN PARAMETERS

For a successful design of a Darrieus-type VAWT, the associated design parameters should be identified and then they should be critically analyzed. These design parameters have close connections with the aerodynamic challenges discussed in the previous section. Islam et al. [10] have identified 22 parameters related to the design of a fixed-pitch SB-VAWT, which are: (i) blade shape, (ii) number of blades, (iii) supporting strut type and shape, (iv) central column, (v) swept area, (vi) solidity, (vii) aspect ratio, (viii) chord/radius ratio, (ix) rated power output, (x) rated wind speed, (xi) cut-in speed, (xii) cut-out speed, (xiii) power coefficient, (xiv) tip speed ratio, (xv) rotational speed, (xvi) pitching of blade, (xvii) tower, (xviii) braking mechanism, (xix) load, (xx) material, (xxi) noise, and (xxii) aesthetics.

It was commented by Islam et al. [10] that—"*It should be mentioned that not all of the parameters are of identical importance for a successful final product. Some of the parameters (like choice of airfoil, supporting strut configuration and shape, solidity, material) are more sensitive and critical than others.*"

In recent years, significant research activities have been carried out with many of the design parameters associated with Darrieus-type VAWTs. In a later section, some of the alternative or innovative design concepts are discussed briefly.

One of the main barriers that is hindering widespread deployment of smaller-capacity Darrieus-type VAWTs is the problem of self-starting, which is closely related to the aerodynamics of VAWTs. Several researchers [9,11–17] have investigated the problem of self-starting through diversified means. Several new designs have attempted to solve the problem of self-starting through design-optimization techniques as discussed in the next sections.

3.3.2 DESIGN OPTIMIZATION

Over the years, researchers have attempted to modify the original two types of Darrieus VAWT configurations (namely eggbeater and straight-bladed) to achieve diversified objectives, including: (i) enhancing the performance, (ii) improving the structural dynamics, (iii) incorporating self-starting ability, (iv) reducing the oscillating torque generation, and (v) improving economic viability. To optimize different design features of a VAWT, different types of aerodynamic models, and experimentation or computational (Computational Fluid Dynamic [CFD], Finite Element Analysis [FEA]) methods are adopted.

3.3.2.1 Aerodynamic Models

One of the key elements of successful VAWT design is an appropriate computational model for performance analysis of Darrieus-type VAWTs. Over the years, many such models have been developed by researchers from different parts of the world. A review of the major computational models can be found in Ref. [18]. It should be noted that these computational models must incorporate different aerodynamics aspects related to VAWTs. A detailed investigation of the main aerodynamic challenges encountered by a SB-VAWT was performed by Islam et al. [19]. According to Islam [20], the key components of all the computational models can be broadly described as:

- "*Calculations of local relative velocities and angles of attack at different tip speed ratios and azimuthal (orbital) positions;*
- *Calculation of the ratio of induced to free stream velocity* $\left(\dfrac{V_a}{V_x}\right)$ *considering the blade/blade–wake interaction;*
- *Mathematical expressions based on different approaches (Momentum, Vortex, or Cascade principles) to calculate normal and tangential forces;*
- *'Pre-stall airfoil characteristics'* (C_l, C_d, *and* C_m) *for the attached regime at different Reynolds numbers;*

- '*Post-stall model*' for stall development and fully stalled regimes;
- '*Finite aspect ratio consideration*';
- '*Dynamic Stall Model*' to account for the unsteady effects;
- '*Flow Curvature Model*' to consider the circular blade motion".

3.3.2.2 Experimental Investigation

Since the invention of the modern Darrieus-type VAWT, researchers in different parts of the world have conducted tests with diversified types of VAWTs for many reasons including: (i) validation of the mathematical or CFD modeling; (ii) improvement of the performance of different VAWTs; and (iii) assessment of the performance of the prototype at in situ situations. Different types of novel instruments (like particle image velocimetry (PIV), hot-wire anemometry, laser Doppler velocimetry, and flow visualizations) have been utilized for investigating different parameters that can give more insights about the operational characteristics of different types of VAWTs. Some of the noteworthy parameters that have been investigated in the past are:

- Performance [21–26]
- Lift, drag, and moment coefficients for two blades [27]
- Blade forces [28]
- Instantaneous pressure distribution [29]
- Hotwire measurements [30]
- Flow visualization [30]
- Thrust [30]
- Blade pitch [31]
- Blade mount point offset [31]
- Visualization of dynamic stall by PIV
- Effect of attachment [32,33]
- Self-starting behavior [16,34]
- Zero-net mass flux actuation [35]
- Aerodynamic loading behavior [36]
- Unsteady wind performance [37]

3.3.2.3 Computational Methods

Among the computational methods, increasingly, designers of VAWTs are utilizing CFD tools. Recently, Untaroiu et al. [38] conducted CFD analysis of a VAWT and they have expressed their optimism in the last paragraph in the conclusion about the prospect of utilizing the CFD analysis. VAWT-related researchers and designers are exploring CFD techniques for investigating numerous parameters, including:

- Flow analysis [39]
- Dynamic stall [40]
- Aerodynamics of the performance [21]
- Performance analysis [21,41]
- Self-starting characteristics [15]
- Unsteady analysis [21,42,43]
- Torque coefficient [44]
- Flow field and power coefficient [45]
- Aerodynamic performance [46]
- Pressure distribution [47]
- Lift and drag coefficients [48]
- Tangential and normal force coefficients [48]
- Power curve [49]

3.3.2.4 Alternative or Innovative Vertical-Axis Wind Turbine Designs

As discussed in the previous section, researchers have different tools to optimize the design of a Darrieus-type VAWT to achieve different objectives. Even through a cursory look into the available literature, one can find numerous design concepts and some of them can be quite different from the original designs of Darrieus [1]. In this section, some of the noteworthy major design improvements are presented.

3.3.2.5 Vertical-Axis Wind Turbines with Alternative Airfoils

The conventional Darrieus-type VAWTs are designed with symmetric airfoils (mainly National Advisory Committee for Aeronautics [NACA] four-digit series), which are infamous for lack of self-starting capability. One of the ways to improve the self-starting capability of conventional Darrieus-type VAWTs is to utilize alternative asymmetric airfoils [20,34]. Recently, Bianchini et al. [13] conducted detailed analysis on start-up behavior of a three-bladed H-Darrieus VAWT through experimental and numerical approaches. It has been found from their analysis that lightly cambered airfoils can indeed provide a significant increase in the starting-torque and impressive reduction of starting times with respect to a symmetric airfoil [16]. However, the overall power production at design point of the slightly cambered airfoil was found to be lower from their analysis.

An attempt was made by Islam et al. [50] to identify different desirable features of an airfoil that are suitable for smaller-capacity fixed-pitch SB-VAWTs. Subsequently, a new airfoil was proposed by Islam et al. [51] that is suitable for smaller-capacity SB-VAWTs. Recently, several other researchers [52–55] have investigated alternative airfoils that are suitable for VAWTs.

3.3.2.6 Helically Twisted Blade

Recently, several researchers have attempted to utilize helically twisted blades to overcome challenges encountered by curved or straight-bladed VAWTs. The main advantage of such blades is reduced flow separation [41,56]. According to Bhutta et al.—*"The resulting rotor thus has a positive lift at zero angle of incidence enabling it to self-start at favorable wind conditions (where ordinary Darrieus rotor may require auxiliary mechanism for starting)"* [5]. However, reported C_p values were as low as 0.128, which is a major disadvantage for twisted blades. Some researchers have also investigated twisted canted blades for VAWTs [25,57]. Based on literature survey, there are at least three commercial VAWT brands that have helical blades [58–60]. For further information on advantages and disadvantages of VAWTs, the interested reader can refer to Section 3 of Ref. [44].

3.3.2.7 Hybrid Darrieus–Savonius Rotor

As discussed earlier, some researchers [61,62] have conducted analysis with hybrid Darrieus- and Savonius-type VAWTs to alleviate the problem of self-starting. Wakui et al. [62] have performed analyses with two types of Darrieus–Savonius hybrid VAWTs and concluded that there is net power extraction for Darrieus–Savonius rotor configuration depending on the wind condition (blowing duration) and the system scale.

3.3.2.8 Darrieus–Masgrowe or Orthogonal Rotor

Gorelov and Krivospitsky [63] investigated a wind turbine designed with the orthogonal rotor of the Darrieus–Masgrowe type (shown in Figure 2 of their article). They analyzed a two-tier SB-VAWT and commented that this type of turbine does not need any devices for the rotor orientation. Furthermore, they can be started independently with a reasonable design. Gorelov and Krivospitsky [63] concluded in their article that their investigation showed a promising future for Darrieus–Masgrowe-type wind turbines. However, further studies are needed to validate this type of wind turbine before wide-scale application.

Apart from the aforementioned designs, several other researchers have investigated some other VAWT configurations. Some of the notable research works with different alternative or innovative concepts are:

- Crossflex concept [64]
- Guide vanes [65–69]
- Circulation control [70,71]
- Lift augmentation [72]
- Magnetic levitated bearing system for VAWTs [73]
- Plasma actuators [74]

3.4 PRACTICAL DARRIEUS VERTICAL-AXIS WIND TURBINE DESIGN METHODOLOGY

Performance analysis of VAWTs requires the characteristics of airfoils in terms of lift, drag, and moment coefficients. In this section, mainly the formulas associated with SB-VAWT-type Darrieus turbines are discussed. It should be noted that the flow velocities in the upstream and downstream sides of the Darrieus-type VAWTs are not constant as shown in Figure 3.3. The chordal velocity component (V_c) and the normal velocity component (V_n) can be calculated from the following expressions

$$V_c = R\omega + V_a \cos\theta \tag{3.1}$$

$$V_n = V_a \sin\theta \tag{3.2}$$

where R is the radius of the turbine, θ is the azimuth angle, ω is the rotational velocity, and V_a is the induced or axial flow velocity through the rotor. The angle of attack (α) can be expressed as

$$\alpha = \tan^{-1}\left(\frac{V_n}{V_c}\right) \tag{3.3}$$

By substituting the values of V_n and V_c and after non-dimensionalizing, we can get

$$\alpha = \tan^{-1}\left[\frac{\sin\theta}{\left(\frac{R\omega}{V_{fs}}/\frac{V_a}{V_{fs}}\right) + \cos\theta}\right] \quad (3.4)$$

Where V_{fs} is known as the freestream wind velocity. If the blade pitching is considered, then the following equation should be used.

$$\alpha = \tan^{-1}\left[\frac{\sin\theta}{\left(\frac{R\omega}{V_{fs}}/\frac{V_a}{V_{fs}}\right) + \cos\theta}\right] - \gamma \quad (3.5)$$

where λ is the blade pitch angle.

The relative flow velocity (W) is calculated as,

$$W = \sqrt{V_c^2 + V_n^2} \quad (3.6)$$

Inserting the values of V_c and V_n from Equations 3.1 and 3.2 into Equation 3.6, the velocity ratio can be expressed as

$$\frac{W}{V_{fs}} = \frac{W}{V_a}\cdot\frac{V_a}{V_{fs}} = \frac{V_a}{V_{fs}}\sqrt{\left[\left(\frac{R\omega}{V_{fs}}/\frac{V_a}{V_{fs}}\right) + \cos\theta\right]^2 + \sin^2\theta} \quad (3.7)$$

In Figure 3.4, the directions of the lift and drag forces and their normal and tangential components are illustrated. The expressions of tangential force coefficient (C_t) and normal force coefficient (C_n) can be written as,

$$C_t = C_l \sin\alpha - C_d \cos\alpha \quad (3.8)$$

$$C_n = C_l \cos\alpha + C_d \sin\alpha \quad (3.9)$$

The net tangential and normal forces are defined as

$$F_t = C_t \tfrac{1}{2}\rho c H W^2 \quad (3.10)$$

$$F_n = C_n \tfrac{1}{2}\rho c H W^2 \quad (3.11)$$

DESIGN OF DARRIEUS-TYPE WIND TURBINES • 55

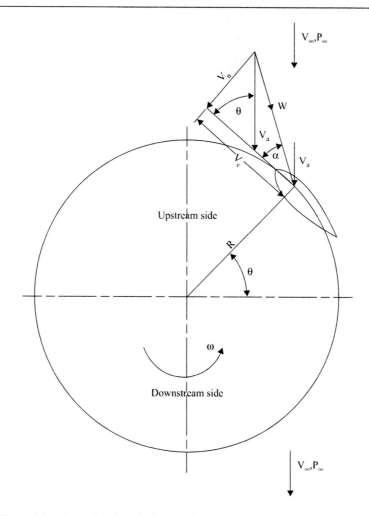

Figure 3.3. Flow velocities of SB-VAWT.

where ρ is the air density, c is the blade chord, and H is the height of the turbine.

The average tangential force (F_{ta}) on one blade can be calculated by

$$F_{ta} = \frac{1}{2\pi} \int_0^{2\pi} F_t(\theta) d\theta \qquad (3.12)$$

The total torque (T) for the number of blades (N) can then be obtained as

$$T = NF_{ta}R \qquad (3.13)$$

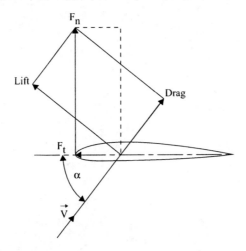

Figure 3.4. Force diagram for a blade airfoil.

Finally, the total power (P) can be calculated from the following expression.

$$P = T \cdot \omega \qquad (3.14)$$

All these formulas are utilized by different types of computational models that should be incorporated into the performance and design analysis of VAWTs.

As stated by Islam [20],

> According to the literature survey, the most studied and best validated models can be broadly classified into three categories—(1) Momentum model, (2) Vortex model, and (3) Cascade model. It should be noted that not all the models consider all the key components described above.

Interested readers can obtain more information about these computational models from Ref. [18].

3.5 CONCLUDING REMARKS

In this chapter, an attempt has been made to deliver a brief overview of Darrieus-type VAWTs and associated design aspects and of the extensive Darrieus literature. It has been demonstrated that the original VAWTs (namely eggbeater and straight-bladed) have gone through diversified modifications over the years. In recent times, several researchers have suggested the application of VAWTs with helically twisted blades, especially for the urban areas and already there are three companies that are manufacturing and marketing this innovative type of VAWT. For design optimization, increasingly researchers are utilizing CFD techniques along with conventional experimental investigation. However, the author was unable to find any updated long-term performance datasets for VAWTs operating in typical real-life situations. Availability of such data would provide important insights about the real-life operational characteristics of VAWTs. At the end, it is anticipated that a self-starting economically designed VAWT is expected to have significant prospect in the coming years.

3.6 ACKNOWLEDGMENTS

The writing of this paper was partly supported by the Canadian Natural Science and Engineering Research Council (NSERC) and the ENMAX Corporation under the Industrial Research Chairs program. The author would like to thank Professor Dr. David Wood of University of Calgary for his encouragement and supports for writing this chapter. His corrections and valuable comments about the draft manuscript of this chapter are also acknowledged.

REFERENCES

[1]. Darrieus, G.J.M. (1931). Turbine having its rotating shaft transverse to the flow of the current. USA, US Patent number 1,835,018. Retrieved from http://patft1.uspto.gov/netacgi/nph-Parser?Sect1=PTO1&Sect2=HITOFF&d=PALL&p=1&u=%2Fnetahtml%2FPTO%2Fsrchnum.htm&r=1&f=G&l=50&s1=1835018.PN.&OS=PN/1835018&RS=PN/1835018.

[2]. Templin, R.J., & Rangi, R.S. (1983). Vertical-axis wind turbine development in Canada, *IEE Proceedings A (Physical Science, Measurement and Instrumentation, Management and Education, Reviews)*, 130(9), 555–561.

[3]. Bhutta, M.M.A., Hayat, N., Farooq, A.U., Ali, Z., Jamil, S.R., & Hussain, Z. (2012). Vertical axis wind turbine—A review of various configurations and design techniques, *Renewable and Sustainable Energy Reviews*, 16(4), 1926–1939. doi:10.1016/j.rser.2011.12.004.

[4]. Eriksson, S., Bernhoff, H., & Leijon, M. (2008). Evaluation of different turbine concepts for wind power, *Renewable and Sustainable Energy Reviews*, 12(5), 1419–1434.

[5]. Islam, M.R., Mekhilef, S., & Saidur, R. (2013). Progress and recent trends of wind energy technology, *Renewable and Sustainable Energy Reviews*, 21, 456–468. doi:10.1016/j.rser.2013.01.007.

[6]. Islam, M., Ting, D.S.-K., & Fartaj, A. (2007a). Assessment of the small-capacity straight-bladed VAWT for sustainable development of Canada, *International Journal of Environmental Studies*, 64(4), 489–500.

[7]. Riegler, H. (2003). HAWT versus VAWT—Small VAWT's find a clear niche, *REFOCUS*, 4(4), 44–46. doi:10.1016/S1471-0846(03)00433-5.

[8]. MacPhee, D., & Beyene, A. (2012). Recent advances in rotor design of vertical axis wind turbines, *Wind Engineering*, 36(6), 647–666. doi:10.1260/0309-524X.36.6.647.

[9]. Kirke, B. K. (1998). *Evaluation of Self-starting Vertical Axis Wind Turbines for Stand-alone Applications*. Australia: Griffith University. Retrieved from Evaluation of Self-Starting Vertical Axis Wind Turbines for Stand-Alone Applications.

[10]. Islam, M., Fartaj, A., & Carriveau, R. (2008). Analysis of the design parameters related to a fixed-pitch straight-bladed vertical axis wind turbine, *Wind Engineering*, 32, 491–507. doi:10.1260/030952408786411903.

[11]. Baker, J. (1983). Features to aid or enable self starting of fixed pitch low solidity vertical axis wind turbines, *Journal of Wind Engineering and Industrial*, 15, 369–380. Retrieved from http://linkinghub.elsevier.com/retrieve/pii/0167610583902064.

[12]. Beri, H., & Yao, Y. (2011a). Numerical simulation of unsteady flow to show self-starting of vertical axis wind turbine using fluent, *Journal of Applied Sciences*, 11(6), 962–970. doi:10.3923/jas.2011.962.970.

[13]. Bianchini, A., Ferrari, L., & Magnani, S. (2011). Start-up behavior of a three-bladed h-Darrieus VAWT: Experimental and numerical analysis, *ASME Conference Proceedings*, Paper number GT2011-45882 (811–820). doi:10.1115/GT2011-45882.

[14]. Hill, N., Dominy, R., Ingram, G., & Dominy, J. (2009). Darrieus turbines: The physics of self-starting, *Proceedings of the Institution of Mechanical Engineers, Part A: Journal of Power and Energy*, 223(1), 21–29. doi:10.1243/09576509JPE615.

[15]. Tanaka, F., Kawaguchi, K., Sugimoto, S., & Tomioka, M. (2011). Influence of wing section and wing setting angle on the starting performance of a darrieus wind turbine with straight wings, *Journal of Environment and Engineering*, 6(2), 302–315.

[16]. Untaroiu, A., Wood, H. G., Allaire, P. E., & Ribando, R. J. (2011a). Investigation of self-starting capability of vertical axis wind turbines using a computational fluid dynamics approach, *Journal of Solar Energy Engineering, Transactions of the ASME*, 133(4). paper number 041010. doi:10.1115/1.4004705.

[17]. Watson, G.R. (1979). The self starting capabilities of low solidity fixed pitch Darrieus rotor, *Proceedings of the 1st British Wind Energy Association Workshop*, 32–39. London: Multi-Science Publishing Co., Ltd.

[18]. Islam, M., Ting, D., & Fartaj, A. (2008). Aerodynamic models for Darrieus-type straight-bladed vertical axis wind turbines. *Renewable and Sustainable Energy Reviews*, 12(4), 1087–1109. doi:10.1016/j.rser.2006.10.023.

[19]. Islam, M., Ting, D.S.-K., Amin, M.R., & Fartaj, A. (2008). Aerodynamic factors affecting performance of straight-bladed vertical axis wind turbines. *ASME International Mechanical Engineering Congress and Exposition*, 6, 331–340. Retrieved from http://www.scopus.com/inward/record.url?eid=2-s2.0-44249085918&partnerID=40&md5=b8ed10d4b-826ecd91b90d41888438ef7.

[20]. Islam, M. (2008). *Analysis of Fixed-Pitch Straight-Bladed VAWT with Asymmetric Airfoils*, PhD Thesis, University of Windsor, Canada.

[21]. Armstrong, S., & Tullis, S. (2011). Power performance of canted blades for a vertical axis wind turbine. *Journal of Renewable and Sustainable Energy*, 3(1), 013106. doi:10.1063/1.3549153.

[22]. Bravo, R., Tullis, S., & Ziada, S. (2007). Performance testing of a small vertical-axis wind turbine. *21st Canadian Congress of Applied Mechanics*. Retrieved from http://www.eng.mcmaster.ca/~stullis/index_files/Bravo CANCAM 2007.pdf.

[23]. Decleyre, W., Van Aerschot, D., & Hirsch, C. (1981). Effects of Reynolds number on the performance characteristic of Darrieus windmills with troposkien and straight blades. BT—*Proceedings of the International Colloquium on Wind Energy*. Organized in conjunction with the International Solar Energy Society, 243–906085594.

[24]. El-Samanoudy, M., Ghorab, A.A.E., & Youssef, S.Z. (2010). Effect of some design parameters on the performance of a Giromill vertical axis wind turbine. *Ain Shams Engineering Journal*, 1(1), 85–95. doi:http://dx.doi.org/10.1016/j.asej.2010.09.012.

[25]. Howell, R., Qin, N., Edwards, J., & Durrani, N. (2010). Wind tunnel and numerical study of a small vertical axis wind turbine. *Renewable Energy*, 35(2), 412–422. doi:http://dx.doi.org/10.1016/j.renene.2009.07.025

[26]. Mertens, S., Van Kuik, G., & Van Bussel, G. (2003). Performance of an H-Darrieus in the skewed flow on a roof. *Journal of Solar Energy Engineering, Transactions of the ASME*, 125(4), 433–440.

[27]. Migliore, P. G., Wolfe, W. P., & Walters, R. E. (1983). Aerodynamic tests of Darrieus wind turbine blades. *Journal of Energy*, 7(2), 101–106.

[28]. Vittecoq, P., & Laneville, A. (1983). Aerodynamic forces for a Darrieus rotor with straight blades: Wind tunnel measurements. *Journal of Wind Engineering and Industrial Aerodynamics*, 15(1–3), 381–388.

[29]. Zervos, A., & Morfiadakis, E. (1990). Instantaneous pressure distribution measurements on the blades of a vertical axis wind turbine (pp. 247–51 BN – 0 9510271 8 2). Laboratory of Aerodynamics, National Technical University of Athens, Greece BT—European Community Wind Energy Conference. Proceedings of an International Conference (EUR 13251), September 10–14, 1990. H.S. Stephens.

[30]. Ferreira, C. J. S., Bussel, G. J. W. van, & Kuik, G. A. M. van. (2006). Wind tunnel hotwire measurements, flow visualization and thrust measurement of a VAWT in skew, *Journal of Solar Energy Engineering*, 128, 487–497. Retrieved from http://link.aip.org/link/?JSEEDO/128/487/1.

[31]. Fiedler, A. J., & Tullis, S. (2009). Blade offset and pitch effects on a high solidity vertical axis wind turbine, *Wind Engineering*, 33(3), 237–246. Retrieved from http://www.swetswise.com.ezproxy.lib.ucalgary.ca/swocweb/linkingDetails.html?openURL=false&issn=0309-524X&eissn=0309-524X&volume=33&issue=3&page=237.

[32]. Li, Y., Tagawa, K., & Liu, W. (2010). Performance effects of attachment on blade on a straight-bladed vertical axis wind turbine, *Current Applied Physics*, 10(2 Suppl.), S335–S338. doi:10.1016/j.cap.2009.11.072.

[33]. Li, Y., & Tagawa, K. (2009). Influence of blade attachment on performance of the straight-bladed vertical axis wind turbine, *Dongli Gongcheng/Power Engineering*, 29(3), 292–296. Retrieved from http://www.scopus.com/inward/record.url?eid=2-s2.0-63549122408&partnerID=40&md5=5e477cbe6bf4f26c2a5ff347573f4812.

[34]. Chinchilla, R., Guccione, S., & Tillman, J. (2011). Wind power technologies: A need for research and development in improving VAWT's airfoil characteristics, *Journal of Industrial Technology*, 27(1). Retrieved from http://www.scopus.com/inward/record.url?eid=2-s2.0-79551587321&partnerID=40&md5=56a4ecd5f0068065613c9e605baa5988.

[35]. Yen, J., & Ahmed, N. (2012). Improving safety and performance of small-scale vertical axis wind turbines, *Procedia Engineering*, 49, 99–106. doi:http://dx.doi.org/10.1016/j.proeng.2012.10.117.

[36]. McLaren, K., Tullis, S., & Ziada, S. (2012). Measurement of high solidity vertical axis wind turbine aerodynamic loads under high vibration response conditions, *Journal of Fluids and Structures*, 32(0), 12–26. doi:http://dx.doi.org/10.1016/j.jfluidstructs.2012.01.001.

[37]. Danao, L.A., & Howell, R. (2012). Effects on the performance of vertical axis wind turbines with unsteady wind inflow: A numerical study, *Methods*, 1–9.

[38]. Untaroiu, A., Wood, H.G., Allaire, P.E., & Ribando, R.J. (2011b). Investigation of self-starting capability of vertical axis wind turbines using a computational fluid dynamics approach, *Journal of Solar Energy Engineering-Transactions of the ASME*, 133(4), paper number 041010. doi:10.1115/1.4004705.

[39]. Horiuchi, K., Ushiyama, I., & Seki, K. (2005). Straight wing vertical axis wind turbines: A flow analysis, *Wind Engineering*, 29, 243–252. doi:10.1260/030952405774354840.

[40]. Ferreira, C.S., & van Bussel, G. (2007). 2D CFD simulation of dynamic stall on a vertical axis wind turbine: Verification and validation with PIV measurements. In *45th AIAA Aerospace Sciences Meeting*, January 8–11, Reno, NV 8–11.

[41]. Gupta, R., & Biswas, A. (2010). Computational fluid dynamics analysis of a twisted three-bladed H-Darrieus rotor, *Journal of Renewable and Sustainable Energy*, 2, 43111. doi:10.1063/1.3483487.

[42]. Danao, L.A., & Howell, R. J. (2012). Effects on the performance of vertical axis wind turbines with unsteady wind inflow: A numerical study, *50th AIAA Aerospace Sciences Meeting including the New Horizons Forum and Aerospace Exposition*, Nashville, TN.

[43]. Qin, N., Howell, R. J., Durrani, N., Hamada, K., & Smith, T. (2011). Unsteady flow simulation and dynamic stall behaviour of vertical axis wind turbine blades, *Wind Engineering*, 35, 511–528. doi:10.1260/0309-524X.35.4.511.

[44]. Mohamed, M.H. (2012). Performance investigation of H-rotor Darrieus turbine with new airfoil shapes. *Energy*, 47(1), 522–530. doi:10.1016/j.energy.2012.08.044.

[45]. Almohammadi, K., Ingham, D., Ma, L., & Pourkashanian, M. (2012). CFD sensitivity analysis of a straight-blade vertical axis wind turbine, *Wind Engineering*, 36(5), 571–588. doi:10.1260/0309-524X.36.5.571.

[46]. Wang, H., Wang, J., Yao, J., Yuan, W., & Cao, L. (2012). Analysis on the aerodynamic performance of vertical axis wind turbine subjected to the change of wind velocity, *Procedia Engineering*, 31, 213–219. doi:http://dx.doi.org/10.1016/j.proeng.2012.01.1014.

[47]. Rolland, S., Newton, W., Williams, A.J., Croft, T.N., Gethin, D.T., & Cross, M. (2013). Simulations technique for the design of a vertical axis wind turbine device with experimental validation, *Applied Energy*, 111, 1195–1203. doi:http://dx.doi.org/10.1016/j.apenergy.2013.04.026.

[48]. Li, C., Zhu, S., Xu, Y.-L., & Xiao, Y. (2013). 2.5D large eddy simulation of vertical axis wind turbine in consideration of high angle of attack flow, *Renewable Energy*, 51, 317–330. doi:10.1016/j.renene.2012.09.011.

[49]. Almohammadi, K.M., Ingham, D.B., Ma, L., & Pourkashan, M. (2013). Computational fluid dynamics (CFD) mesh independency techniques for a straight blade vertical axis wind turbine, *Energy*, 58, 483–493. doi:10.1016/j.energy.2013.06.012.

[50]. Islam, M., Ting, D.S.-K., & Fartaj, A. (2007b). Desirable airfoil features for smaller-capacity straight-bladed VAWT, *Wind Engineering*, 31(3), 165–196. doi:10.1260/030952407781998800.

[51]. Islam, M., Ting, D.S.-K., & Fartaj, A. (2007c). Design of a special-purpose airfoil for smaller-capacity straight-bladed VAWT, *Wind Engineering*, 31(6), 401–424. doi:10.1260/030952407784079780.

[52]. Bah, E.A.A., Sankar, L.N., & Jagoda, J. (2013). Investigation on the use of multi-element airfoils for improving vertical axis wind turbine performance. In *51st AIAA Aerospace Sciences Meeting Including the New Horizons Forum and Aerospace Exposition 2013* (Vol. Grapevine), Atlanta, GA. Retrieved from http://www.scopus.com/inward/record.url?eid=2-s2.0-84881462465&partnerID=40&md5=149e2289f1f813798afce652cae6c58e.

[53]. Beri, H., & Yao, Y. (2011b). Effect of camber airfoil on self starting of vertical axis wind turbine, *Journal of Environmental Science and Technology*, 4(3), 302–312. doi:10.3923/jest.2011.302.312.

[54]. Castelli, M. R., Grandi, G., & Benini, E. (2012). Numerical analysis of the performance of the du91-w2-250 airfoil for straight-bladed vertical-axis

wind turbine application, *International Journal of Mechanical and Aerospace Engineering*, 6, 329–334.

[55]. Saeed, F., Paraschivoiu, I., Trifu, O., Hess, M., & Gabrys, C., (2008). Inverse airfoil design method for low-speed straight-bladed Darrieus-type VAWT applications, *7th World Wind Energy Conference*, Kingston, ON, Canada. Retrieved from https://eprints.kfupm.edu.sa/1286/.

[56]. Gupta, R., Sukanta, R., & Biswas, A. (2010). Computational fluid dynamics analysis of a twisted airfoil shaped two-bladed H-Darrieus rotor made from fibreglass reinforced plastic (FRP), *International Journal of Energy and Environment*, 1, 953–968.

[57]. Armstrong, S., Fiedler, A., & Tullis, S. (2012). Flow separation on a high Reynolds number, high solidity vertical axis wind turbine with straight and canted blades and canted blades with fences, *Renewable Energy*, 41, 13–22. doi:http://dx.doi.org/10.1016/j.renene.2011.09.002.

[58]. Quietrevolution. (2013). Retrieved November 12, 2013, from http://www.quietrevolution.com/qr5/qr5-turbine.htm.

[59]. Turby. (2013). Retrieved November 12, 2013, from http://www.turby.nl

[60]. UGE. (2013). Retrieved November 12, 2013, from http://www.urbangreenenergy.com.

[61]. Elmabrok, A.M. (2009). Estimation of the performance of the Darrieus Savonius combined machine, *International Conference and Exhibition on Ecological Vehicles and Renewable Energies*, March 26–29, Monaco. Retrieved from http://cmrt.centrale-marseille.fr/cpi/ever09/documents/papers/re7/EVER09-paper-77.pdf.

[62]. Wakui, T., Tanzawa, Y., Hashizume, T., & Nagao, T. (2005). Hybrid configuration of Darrieus and Savonius rotors for stand-alone wind turbine-generator systems, *Electrical Engineering in Japan*, 150(4), 13–22. doi:10.1002/eej.20071.

[63]. Gorelov, D.N., & Krivospitsky, V.P. (2008). Prospects for development of wind turbines with orthogonal rotor, *Thermophysics and Aeromechanics*, 15, 153–157. doi:10.1134/S0869864308010149.

[64]. Sharpe, T., & Proven, G. (2010). Crossflex: Concept and early development of a true building integrated wind turbine, *Energy and Buildings*, 42(12), 2365–2375. Retrieved from http://www.sciencedirect.com/science/article/pii/S0378778810002653.

[65]. Wu, G.Q., Chen, X., Cao, Y., & Zhou, J.L. (2010). Simulation and test for two airfoils with wind guide vane of VAWT, *Advanced Materials Research*, 148–149, 1199–1203. doi:10.4028/www.scientific.net/AMR.148-149.1199.

[66]. Chong, W.T., Fazlizan, A., Poh, S.C., Pan, K.C., Hew, W.P., & Hsiao, F.B. (2013). The design, simulation and testing of an urban vertical axis wind turbine with the omni-direction-guide-vane, *Applied Energy*, 112, 601–609. doi:http://dx.doi.org/10.1016/j.apenergy.2012.12.064.

[67]. Takao, M., Maeda, T., Kamada, Y., Oki, M., & Kuma, H. (2008). A straight-bladed vertical axis wind turbine with a directed guide vane row, *Journal of Fluid Science and Technology*, 3(3), 379–386. doi:10.1299/jfst.3.379.

[68]. Kuma, H., Takao, M., Beppu, T., Maeda, T., Kamada, Y., Kamemoto, K., & Asme. (2008). A straight-bladed vertical axis wind turbine with a directed guide vane—mechanism of performance improvement, *Proceedings of the 27th International Conference on Offshore Mechanics and Arctic Engineering—2008*, 6, 617–623. doi:10.1115/OMAE2008-57233.

[69]. Takao, M., Takita, H., Maeda, T., & Kamada, Y. (2009). A straight-bladed vertical axis wind turbine with a directed guide vane row—Effect of guide vane solidity on the performance, *19th International Offshore and Polar Engineering Conference*, Osaka, Japan, June 21–26, 2009.

[70]. Wilhelm, J.P., Panther, C.C., Pertl, F.A., & Smith, J.E. (2009). Vortex analytical model of a circulation controlled vertical axis wind turbine, *ES2009-90348*, 1–8. doi:10.1115/ES2009-90348.

[71]. Wilhelm, J.P., Panther, C., Pertl, F.A., & Smith, J.E. (2009). Momentum analytical model of a circulation controlled vertical axis wind turbine, *ASME 3rd International Conference on Energy Sustainability, ES2009*, San Francisco, CA, July 19–23, 2009, paper number ES2009-90352.

[72]. Ii, G.M.A., Pertl, F.A., Clarke, M.A., & Smith, J.E. (2010). Lift augmentation for vertical axis wind turbines, *International Journal of Engineering*, 4, 430–442. Retrieved from http://www.doaj.org/doaj?func=abstract&id=677050.

[73]. Kumbernuss, J., Jian, C., Wang, J., Yang, H.X., Fu, W.N. (2012). A novel magnetic levitated bearing system for vertical axis wind turbines (VAWT), *Applied Energy*, 90(1), 148–153. doi:10.1016/j.apenergy.2011.04.008.

[74]. Greenblatt, D., Schulman, M., & Ben-Harav, A. (2012). Vertical axis wind turbine performance enhancement using plasma actuators, *Renewable Energy*, 37(1), 345–354. doi:10.1016/j.renene.2011.06.040.

CHAPTER 4

DESIGN OF SAVONIUS-STYLE WIND TURBINES

Sukanta Roy, Ujjwal K. Saha

This chapter presents an exhaustive review of Savonius-style wind turbines. These designs spin about a vertical axis and are driven by fluid drag forces. They rotate slower than their lift-drive counterparts and create high torque. Savonius turbines are able to be used in confined spaces and in areas where wind direction changes. They are particularly suited for small-scale power generation.

The chapter details the historical development of Savonius turbines from their inception until modern days. The authors show performance characteristics and provide guidance to optimize the power-generation capability through aspect ratios, blade overlap, blade twist, multiple stages, end plates, and profile. Information here will allow readers to design, construct, and utilize small-scale Savonius blades for a variety of power-production activities.

4.1 INTRODUCTION

The most basic type of vertical-axis wind turbine (VAWT) is the Savonius-style wind turbine. This class of wind turbine was introduced by a Finish engineer named Savonius in the 1920s and was named after its inventor [1]. This is basically a modification of the Flettner's rotor used in ships (large cylinders mounted on the vertical axis). With a Flettner's rotor, the wind pressure difference created across the cylinder by means of the "Magnus effect" (forces generated on curved objects that move through a fluid) was mainly responsible to propel the ships.

The Savonius-style turbine is a simple device, designed by cutting a cylinder into halves, along its central axis and relocating the semi-cylindrical surfaces sideways. The outlook of this turbine is analogous to an "S" when viewed from the top. These turbines are designed to be driven by the wind drag forces on the turbine blades, and in this regard, they differ in appearances from their more common counterparts that are lift-driven VAWTs. The schematic diagrams of "S"-shaped and helical-shaped Savonius-style turbines are shown in Figure 4.1 along with the most important dimensions depicted in Figure 4.2.

The power-conversion capability of these turbines is inferior to that of other wind energy convertors. A typical conventional Savonius-style turbine possesses a power conversion efficiency of 12%–18%, as compared to the efficiency of 30%–40% of vertical-axis Darrieus-style wind turbines. Horizontal-axis wind turbines, on the other hand, give efficiency in the range of 40%–50%. Thus, Savonius-style turbines are not competitive with other styles of wind turbines in terms of their aerodynamic performance, and should only be considered as application-specific devices for the conversion of wind energy, and small-scale technological alternatives to other wind turbines. Nevertheless, it is worth mentioning that with a modified design of Savonius-style turbines, the peak power conversion efficiency can reach up to 30% [2–5].

These turbines have a very low cut-in speed and can operate in winds as low as 3 m/s (6.7 mph). They can operate in a wide range of wind speeds and produce lesser vibrational loads on the supporting structure due to their lower rotational speed. Design simplicity, low cost, and easy

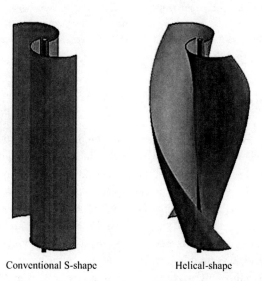

Conventional S-shape   Helical-shape

Figure 4.1. Two-bladed "S"-shaped and helical-shaped Savonius-style vertical-axis wind turbines.

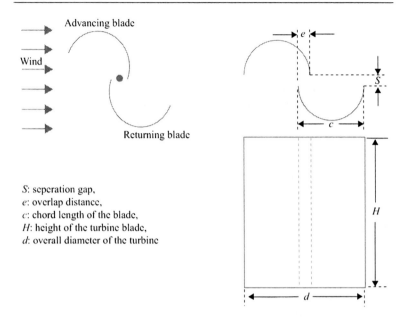

Figure 4.2. Dimensions of a Savonius-style vertical-axis wind turbine.

assembly are the advantages of these turbines. These devices are insensitive to the wind flow directions, and thus, are very useful for the specific locations of variable wind directions. Their vertical rotational axis allows them to be positioned in multiple numbers in a confined space or atop buildings [2–4].

This class of turbine is a very useful device for water pumping in agricultural purposes, particularly, in rural areas where the water level is within a distance of 5 m (16.4 ft) below the ground. The installation and maintenance costs are very low and it can be installed on rooftops for local electricity production. It is also useful in buildings for heating, ventilation, and air-conditioning purposes. A Flettner ventilator, which is often seen on the roofs of vans and buses and is used as a cooling device is another practical application of the Savonius-style VAWTs. Other applications are in decentralized small-scale electricity generation for remote areas. A robust wind turbine system equipped with battery storage capacity can be a viable option for intermittent electricity generation. This stored electricity can be used for running electronic equipment, lighting or charging mobiles. Nowadays, another promising application of these turbines can be observed on cellular communication towers. In a modern communication tower, the usual power requirement is around 1–3 kW for cell-phone electronics associated with it, which is collected either from the grid connections or from a generator system. With the rapid advancement of cellular technology, communication towers with antennas are being built in urban as well as rural areas. In some rural areas where grid connectivity is not available, these small-scale turbines can be a viable option to generate the required power. Moreover, they will generate more power at high altitudes due to higher wind speeds and lesser obstructions. This application gives an additional benefit of reduction in the requirement of the separate base and tower [2–9]. Currently, another application of these turbines is observed in association with the Darrieus-style VAWTs due to better starting ability of the Savonius-style wind turbines. A photograph showing a typical combined Savonius–Darrieus-style turbine is presented in Figure 4.3.

Figure 4.3. Historic Gold Mine Savonius–Darrieus combined small-scale vertical axis turbine in Taiwan (November, 2009).

Source: Courtesy of Fred Hsu, reproduced under GNU Free Documentation License 1.2.

This chapter is intended to discuss in depth the basic principles, design aspects, and performance measures of the Savonius-style wind turbines, focusing on the current status of these turbines.

4.2 WORKING PRINCIPLE OF SAVONIUS-STYLE WIND TURBINES

Savonius-style wind turbines mainly rotate due to exertion of wind drag force between the convex and concave parts of the turbine blades when they rotate around a vertical shaft. However, lift also contributes to the power generation at various rotational angular positions. During the complete rotational cycles of Savonius-style turbines, various types of flow patterns such as free stream, coanda-type, overlap, separation, stagnation, and vortex flows are observed around the turbine blades [3]. Figure 4.4 demonstrates the various types of flow patterns at different angular positions (θ).

The free stream wind flow (I) proceeds to various flow patterns once it acts on the Savonius-style turbines as shown in Figure 4.4. Coanda-type flow (II), primarily accountable for producing lift forces on the turbine blades, is prominent at angular positions in the range of $\theta = 0° - 45°$. The pressure drag on the convex sides of returning blades at $\theta = 60° - 150°$ is restored by the dragging flow (III) and overlapping flow (IV) on the turbine blades. Flows (III) and (IV) create a returning effect on the concave side of the returning blade, which in turn helps in restoring the negative pressure drag forces acting against the convex side of the returning blade. Thus, flow types (II), (III) and (IV) contribute to the enhancement of averaged power coefficient. Flow (V) separates the attached flow from the turbine blades, and causes the vortex formation (VIII). Formation of vortex shedding is a common phenomenon for the flows around Savonius-style turbines. Vortex shedding effects are formed near the tip of the returning blade up to $\theta < 90°$, beyond which these are separated from the tip and formed at the downstream edge of the turbine blades. The stagnation flow (VI), returning flow (VII), and vortex flow (VIII) are found to reduce the average power of the turbine.

The performance of these turbines is more often articulated in terms of the power coefficient (C_p) and torque coefficient (C_T), and is expressed in the form of Equation 4.1.

$$C_p = \frac{P_{turbine}}{P_{available}} = \frac{T\omega}{\frac{1}{2}\rho A V^3} = \frac{T}{\frac{1}{2}\rho A V^2 r}\frac{r\omega}{V} = C_T \lambda \qquad (4.1)$$

70 • SMALL-SCALE WIND POWER

where,

> T is the moment (N m),
> ρ is the density of the air (kg/m³),
> V is the free stream wind speed (m/s),
> λ is the tip speed ratio (*TSR*),
> r is the radius of rotation of the turbine (m),
> ω is the rotational speed of the wind turbine,
> $P_{turbine}$ is the power produced by the turbine (W), and
> $P_{available}$ is the power available in the wind (W).

The graphical representation of Figures 4.5 and 4.6 are widely used to compare the performance of Savonius-style wind turbines with those of

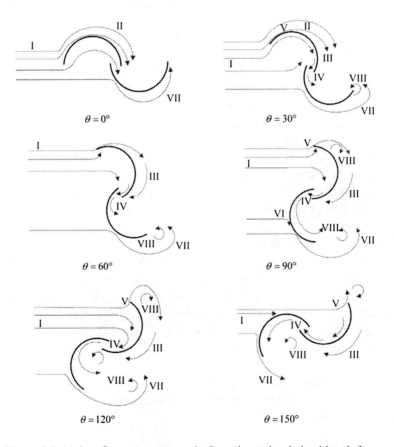

Figure 4.4. Various flow patterns around a Savonius-style wind turbine (I: free stream flow, II: Coanda type flow, III: dragging-type flow, IV: overlapping flow, V: separation flow, VI: stagnation flow, VII: returning flow, VIII: vortex flow).

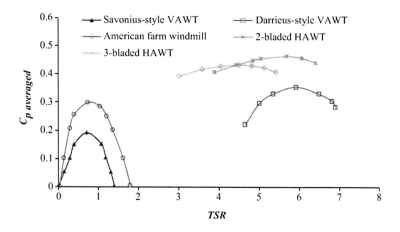

Figure 4.5. Power coefficient of Savonius-style "S"-shaped wind turbines along with other counterparts.

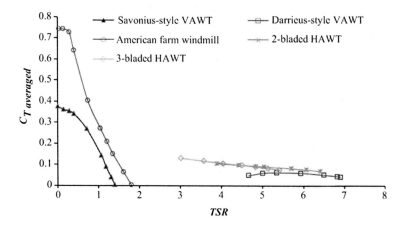

Figure 4.6. Torque coefficient of Savonius-style "S" shaped wind turbines along with other counterparts.

other existing wind turbines showing a lower average power coefficient for conventional Savonius-style turbines. However, a better torque characteristic signifies the ability to operate at low operating wind speeds and easy starting. A performance comparison obtained from wind tunnel tests in terms of maximum power coefficient for conventional Savonius-style wind turbines is demonstrated in Table 4.1.

Table 4.1. Maximum power coefficient of conventional Savonius-style wind turbines

Investigators	$H \times D$ (m × m)	Wind Tunnel Dimension (m × m)	TSR	Max C_P
Alexander and Holownia [10]	0.46 × 0.19	2.4 × 1.2 (closed)	0.52	0.147
Baird and Pender [11]	0.076 × 0.06	0.305 × 0.305 (closed)	0.78	0.185
Bergeles and Athanassiadis [12]	0.7 × 1.4	3.5 × 2.5 (closed)	0.7	0.128
Sivasegaram and Sivapalan [13]	0.12 × 0.15	0.46 × 0.46 (open)	0.75	0.20
Bowden and McAleese [14]	0.164 × 0.162	0.76 (circular open)	0.72	0.15
Ogawa and Yoshida [15]	0.175 × 0.3	0.8 × 0.6 (open)	0.86	0.17
Fujisawa and Gotoh [16]	0.32 × 0.32	0.5 × 0.5 (open)	0.9	0.17
Huda et al. [17]	0.185 × 0.32	0.5 (circular open)	0.71	0.17
Hayashi et al. [18]	0.23 × 0.33	1.5 × 1.5 (open)	0.75	0.175 (single stage)
	0.074 × 0.184			0.13 (three stage)
Kamoji et al. [19]	0.208 × 0.208	0.4 × 0.4 (open)	0.78	0.161 (single stage)
	0.226 × 0.113		0.83	0.145 (two stage)
Dobrev and Massouh [20]	0.2 × 0.22	1.35 × 1.65 (closed)	0.8	0.18

4.3 DESIGN PARAMETERS OF SAVONIUS-STYLE TURBINES

In order to design an efficient Savonius-style wind turbine, the most important design parameters are scaling factor, solidity factor of the turbine, turbine height-to-diameter ratio, overlap distance, and separation gap between the blades, turbine end plates, and blade profile. In this section, a detailed description is given, focusing on the current status of this turbine.

4.3.1 SCALING FACTOR

One of the major parameters that affects the performance and design optimization of a Savonius-style turbine is the scaling of large turbines into the small models. In practice, the tested results of small models often suffer from inaccuracy. That is mainly due to the presence of electrical and mechanical losses related to energy conversion from shaft to generator in the large models, and also due to the blade geometry to mass relationships. To maintain the similarity of performance parameters (C_p and C_T), the flow conditions must be the same. Using the theory of similarity, this can be achieved by (i) maintaining the same TSR; (ii) maintaining the blade profile, number of blades used, and the materials used; and (iii) making proportional adjustments to all dimensions like chord length, blade radius, aspect ratio, overlap ratio, and so forth [21].

4.3.2 SOLIDITY FACTOR

The solidity factor (S_f) is an important criterion for designing a wind turbine and is defined as,

$$S_f = \frac{Total\ area\ of\ the\ turbine\ blades}{Swept\ area\ normal\ to\ the\ direction\ of\ wind\ flow} \quad (4.2)$$

In an efficiently designed Savonius-style wind turbine, a comparatively larger area of the turbine blades is intercepted by a small area of wind, while the area is compared with the case of lift-driven turbines. It therefore has a high-solidity factor, which is not desirable for a higher-rotational system. This style of turbines with high solidity factor usually suffers from a high degree of aerodynamic interference between the blades, which results in low values of *TSR* and power coefficient C_p. Consequently, this type of turbine usually operates at low operating speeds but has high-starting

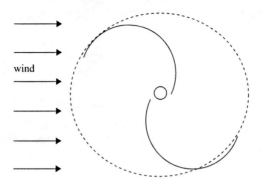

Figure 4.7. Savonius-style turbines.

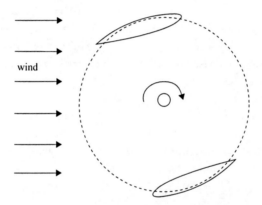

Figure 4.8. Darrieus-style turbines

torque and in this regard, it differs from its counterparts that are lift driven. A better understanding can be drawn from the examples demonstrated in Figures 4.7 and 4.8. The former depicts the solidity of Savonius-style turbine and the latter shows the solidity of its counterpart Darrieus-style turbine, keeping the swept area constant.

4.3.3 ASPECT RATIO

The turbine aspect ratio is derived as a nondimensional parameter through dividing the height of the turbine (H) by its diameter (d). It is a decisive parameter for satisfactory performance of Savonius-style turbines consid-

DESIGN OF SAVONIUS-STYLE WIND TURBINES • 75

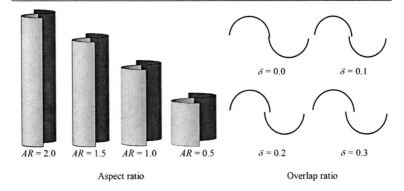

Figure 4.9. Examples of Savonius-style turbines with different aspect ratios and overlap ratios.

ering the effect of solidity factor because of its impact on the performance coefficient (C_p). It is often expressed as:

$$AR = \frac{H}{d} \qquad (4.3)$$

The dimensions of this style of turbine are demonstrated in Figure 4.2 and a clear picture of different aspect ratios can be seen from Figure 4.9. A low aspect ratio (< 2) is often preferred for these turbines to give structural stability. A small diameter always causes sharp turns in the direction of airflow that accelerate the turbine rotational speed. In contrast, with an increase of diameter, the torque generated by the turbine blades increases at the cost of decreased rotational speed. In the present scenario, most of the existing commercialized Savonius-style turbines have an aspect ratio of 1.5–2.0 [4].

4.3.4 OVERLAP RATIO

The overlap ratio of the Savonius-style turbines plays a vital role in optimizing the performance of these turbines. This ratio is obtained by dividing the overlap distance between the turbine blades (e), by the blade chord length (c) and is given by,

$$\delta = \frac{e}{c} \qquad (4.4)$$

An effective design of a conventional Savonius-style turbine with overlapping blades proved to have better starting characteristics compared to that

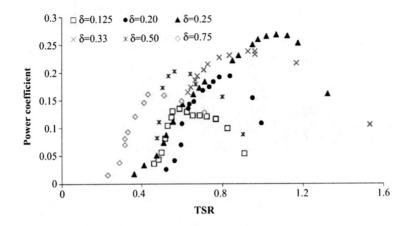

Figure 4.10. Effect of overlap ratio on the performance of Savonius-style wind turbines.

Source: Reproduced with permission from Mojola [22]. © 1985 by Elsevier Ltd.

of a nonoverlapping system. This improvement is mainly triggered by the overlapping flow between the turbine blades. This flow reduces the negative torque produced by the returning blade of the turbine by inserting an opposite effect on the concave side of the returning blade. Hence, the average power of the turbine gets increased. A group of systematic drawings, showing the wind flow through the overlap distance at various rotational positions of the turbine, is given in Figure 4.4. It is obvious that as the overlap ratio increases beyond an optimum value, the effective pressure on the concave side of the advancing blade reduces. Thus, a small overlap distance of ($\delta = 0.15$–0.25) is always preferable in order to have a better performance of these turbines [4,9].

Figure 4.10 shows the effect of overlap ratio on the performance of Savonius-style turbines with respect to TSR, a maximum power coefficient of 0.27 was achieved with an overlap ratio of 0.25.

4.3.5 SEPARATION GAP

The separation gap (s) is shown in Figure 4.2. It is responsible for the airflow through the space between the blades. With the addition of a separation gap, wind cannot focus on the concave portion of the returning blade, and as a result, the advantage of restoring pressure drag on the returning blade through the overlapping flow and dragging flow is much reduced, which in turn reduces the average power of the turbine. Most of the investigations showed a better performance with no separation gap than the design comprising a separation gap [3].

4.3.6 END PLATES

The application of end plates in the turbine design seemed to have enhanced the performance. It is a simple device with a negligible thickness, and is installed as a cap on the top and bottom of the turbine blades, as shown in Figure 4.11. The addition of end plates on a Savonius-style turbine marginally increases the maximum average power coefficient, and also helps in better operation at higher TSRs. It is mainly due to the fact that the end plates inhibit the air outflow from the concave side of the blades, keeping the pressure difference between concave and convex sides of the blades at satisfactory levels over the height of the turbine. The diameter of the end plates is recommended to be 1.1 times the turbine diameter to give a better performance [3–4, 6–9].

4.3.7 MULTI-STAGING

As discussed in the earlier sections, a high starting torque and slow operating wind speed are a major advantage of Savonius-style turbines.

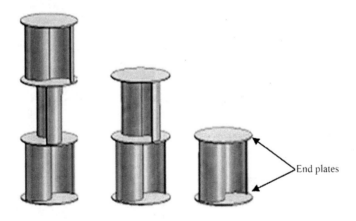

Figure 4.11. Application of end plates and multi-staging in the design of Savonius-style wind turbines.

Source: Adapted by permission from Akwa et al. [3]. © 2012 by Elsevier Ltd.

However, at some angular positions ($\theta = 110° - 160°$), it is observed that changes in the wind direction cause a low starting torque such that the turbine cannot start on its own. To overcome this drawback, the use of multi-staging has shown a significant improvement over the single-stage Savonius-style turbines. A two-stage design of this style is accomplished by setting the upper and lower blade pairs at 90° to each other, and it is 120° to each other in the case of a three-stage design. Figure 4.11 shows the arrangement for different staged systems. The use of multi-staging reduces the high fluctuation of torque without significant performance loss of the turbine, operating with cycles lagged relative to one another [18,23]. Lower fluctuation in torque is illustrated by the graphic representation of Figure 4.12.

4.3.8 NUMBER OF BLADES

The number of blades for Savonius-style wind turbines is an effective parameter depending on the operating conditions. Two-bladed turbines of this style have shown a better performance over their three-bladed counterparts. However, three-bladed turbines have a higher starting torque as compared to two-bladed turbines. This is mainly due to the reduction in angular positions of the advancing blade, resulting in a "cascade effect" in which each blade affects the performance of the following blade. As a consequence, a lesser amount of energy released

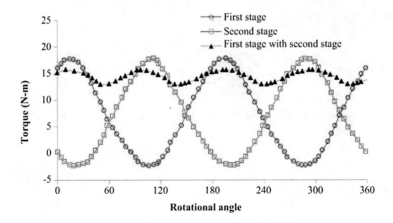

Figure 4.12. Effect of multi-staging on the torque performance of Savonius-style wind turbine (reproduced by permission from Akwa et al. [3]. © 2012 by Elsevier Ltd).

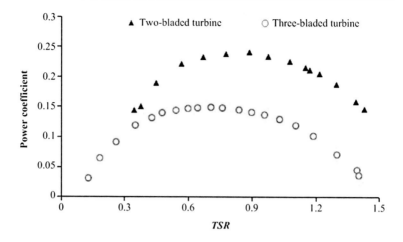

Figure 4.13. Comparative analysis of two- and three-bladed Savonius-style turbines.

Source: Reproduced from Blackwell et al. [24].

by the moving air is converted into mechanical energy by higher-bladed systems. The power coefficient of a two-bladed turbine is about 1.5 times higher than that of the three-bladed turbine. Figure 4.13 shows a comparative analysis of power coefficients for two- and three-bladed Savonius turbines. A better performance is observed with the two-bladed system. However, as discussed earlier, the starting capability can be enhanced for a two-bladed turbine by adapting a multi-staging system.

4.3.9 BLADE PROFILES

The benefit on the use of a simple blade profile for Savonius-style wind turbines gives the flexibility in variable blade profile optimization. The first Savonius-style turbine of "S"-shape was patented by Savonius in 1929 [25]. Thereafter, he modified the design to increase the efficiency of the turbine 1930 [26]. These designs were further modified by Benesh to improve the efficiency of the turbines [27,28]. Figure 4.14 shows the different blade profiles patented by Savonius and Benesh, which are mostly used in Savonius-style turbines.

The blade profile of conventional "S"-shape has been modified from the beginning of its existence [29,30]. An optimized profile as shown in Figure 4.15 claimed a power conversion efficiency of 30%.

Inventor	Patent No.	Blade Profile
Savonius (1929)	US Patent: 1,697,574	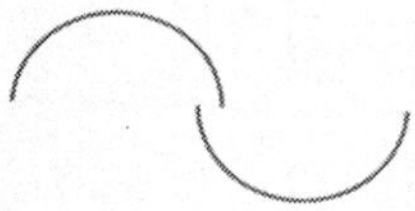
Savonius (1930)	US Patent: 1,766,765	
Benesh (1988)	US Patent: 4,784,568	
Benesh (1996)	US Patent: 5,494,407	

Figure 4.14. Demonstration of most common Savonius-style blade profiles patented by Savonius and Benesh.

Source: Reproduced from patents Savonius and Benesh [25–28].

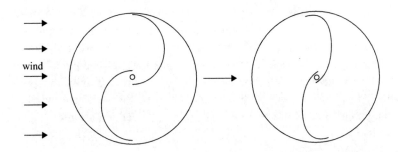

Figure 4.15. Optimized blade profile of Savonius-style turbine.

Source: Reproduced with permission from Mohamed et al. [30]. © 2011 by Elsevier Ltd.

A practical example of these optimized blade profiles of Savonius-style turbines to be used on communication towers is shown in Figure 4.16.

Apart from this "S"- or "hook"-type profiles, the other type of profile that is found in some applications is twisted blades. As reported in open literatures, twisted-bladed turbines have marginally improved the efficiency and starting characteristics of these turbines [23,32,33,34]. A photograph showing the practical implementation of twisted blades, manufactured by Tangarie, is displayed in Figure 4.17. A modification of

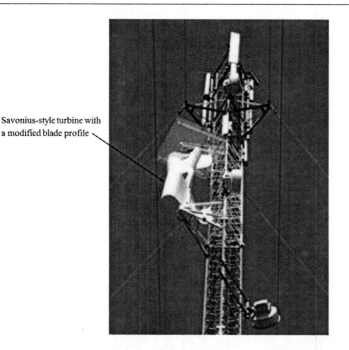

Figure 4.16. Application of Savonius-style wind turbines in communication towers with a modified blade profile.

Source: Adapted with permission from Plourde et al. [31]. © 2011 by International Frequency Sensor Association.

twisted blades is helical blades, which can be visualized as a turbine of many stages with negligible heights that are stacked one upon another in such a fashion that it smoothly tends to yield a twist of 180°. The effect of helical designs reduces the large fluctuations in the torque characteristic of a single-stage Savonius-style turbine; however, the behavior is close to that of adding stages in the conventional turbines. The performance of these helical-shaped turbines does not differ significantly from that of modified "S"- or "hook"-type blades [31]. A photograph showing a 3 kW helical blade turbine is demonstrated in Figure 4.18.

4.4 TESTING AND PERFORMANCE MEASURES

Experimental investigations on Savonius-style wind turbines were carried out either in the wind tunnel or in open fields. Examples of practical utilization in the open field are presented in Figures 4.3, 4.16, 4.17, and 4.18, where by means of generator-motor assembly, the efficiency of these turbines can be

Figure 4.17. A twisted-bladed Savonius-style vertical-axis wind turbine "Gale" by Tangarie, demonstrated at Crissy Field Small VAWT Demonstration, San Francisco (2012).

Source: Photo by Paul Gipe. All rights reserved [35].

Figure 4.18. Photograph of a 3 kW helical vertical-axis wind turbine.

Source: Reproduced with permission from Soluciones Energéticas S. A. [36]. © by manufacturer SolucionesEnergéticas, S.A.

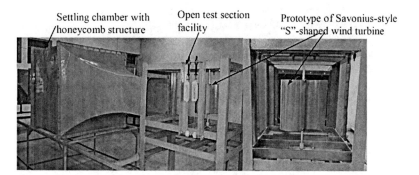

Figure 4.19. Close-up photograph showing testing of conventional Savonius-style turbines in a wind tunnel with open test section facility.

Source: Courtesy of Sukanta Roy, Indian Institute of Technology Guwahati.

calculated through Equation 4.1. However, most of the investigations found in the open literature have shown the use of wind tunnels either with a closed test section inside the wind tunnel or an open-type test section at the exit of the wind tunnel, for testing of prototype turbine models with a proper scale.

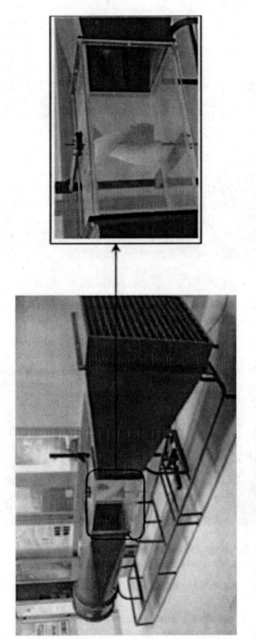

Figure 4.20. Testing of Savonius-style helical wind turbines in a wind tunnel with a closed test section.

Source: Adapted with permission from Damak et al. [38]. © 2012 by Elsevier Ltd.

Visualization of these two types of wind tunnels is demonstrated in Figures 4.19 and 4.20. During the last decade, some low-speed wind tunnel studies were carried out by a host of researchers at IIT Guwahati [7,23,32,37].

Among the issues with the wind tunnel studies, the most critical is the blockage effect, which must be taken into account in the case of closed test section facility inside the wind tunnel. In this type of test sections, the air moves past the turbine, and accelerates due to the conservation-of-mass principle. Since the available wind power varies with the cube of the wind speed, even a small blockage ratio can largely influence the power coefficient of the turbine. However, the determination of an appropriate blockage correction becomes challenging when a complex shape of Savonius style is tested. It can only be determined properly by testing the same model in the open air. It can also be determined by computational fluid dynamics (CFD) and comparing results of simulations in and out of tunnels [6–7]. Since, with a change in the shape and size, the blockage factor will also be changed, a particular blockage factor is only valid for those particular tested designs. The effect of a blockage factor can be seen in Figure 4.21, indicating the significance of blockage corrections (B) of 2%–20% [39].

In contrast, a wind tunnel with an open test section facility sometimes suffers in maintaining the streamlined air flow. Design of the settling chamber that contains a honeycomb structure and the screen is the prime factor to have a streamlined airflow.

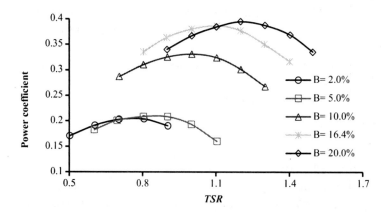

Figure 4.21. Effect of blockage on the performance of Savonius-style wind turbine.

Source: Data points taken from Modi and Fernando (1989) [39]. Copyright 1989 ASME.

Another serious issue that must be minimized is the losses in a turbine assembly system, which broadly includes the frictional losses in the bearings, shafts, and power transmission devices, and the losses incurred in the electrical system meant for power conversion. Other issues related to wind tunnel testing involve the efficient measurement of various parameters, such as wind speed measurement through digital anemometers, turbine rotational speed measurements by optical tachometers, power measurement through generator-motor assembly, pressure measurement by means of pressure sensors, torque measurement through torque transducers or dynamometers, and so forth. A systematic approach to the experimental investigation can lead to a more efficient design of these turbines. However, it is generally recommended to test Savonius-style turbines in an open testing facility.

4.5 MATHEMATICAL AND COMPUTATIONAL MODELS

As can be observed from the preceding sections and from the literature, the flow around a Savonius-style turbine is time dependent and complex in nature; separation and vortex formation are common phenomena. Thus, a reliable theoretical analysis of the flow around a Savonius-style wind turbine is extremely difficult, although not impossible. In contrast, several well-developed theories are available to analyze the performance of lift-based horizontal-axis wind turbines and Darrieus-style wind turbines. The more popular blade element theory, which is discussed in Chapters 3 and 6 of this book, assumes the span-wise blade elements to be independent of each other. The forces on the blade elements can be determined by the local flow properties that are estimated by their momentum and vortex consideration.

However, this method cannot be applied directly to the entirely distinct flow around a Savonius-style turbine. The classical momentum theory also fails to analyze the performance of a Savonius-style turbine accounting the effect of various parameters [39]. The streamtube model, based on the blade element theory, is found effective for the analysis of a Darrieus-style turbine at low TSRs [40]. However, it is argued to be ineffective in predicting the performance of a Savonius-style turbine [41,42]. It is mainly due to the larger turbine area of Savonius-style turbine to the flow swept area. Moreover, the staggered curved blades restrict the streamtube model to form the entire flow field as an isolated cross-sectional flow pattern and also to calculate the forces on the curved surfaces.

For Savonius-style wind turbines, the initial analytical studies considering the vortex shedding have not assumed the effect of flow separation, and as a result, unrealistic performance predictions were observed; however, these formed the basis of many later theoretical studies [43,44]. To deal with the problems of flow separation and vortex formation around these turbines, available literatures show a qualitative potential of the discrete vortex method considering the effect of separation, where the flow is assumed to be a blend of a finite number of discrete vortices [39,45,46]. The velocities and pressure distributions on both sides of the turbine blades are obtained with an assumption of uniformly distributed vortices over the length of the element. The torque and power generated can be obtained from the velocity and pressure distribution.

Alternatively, the recent developments of three-dimensional finite volume–based commercial codes (e.g., ANSYS Fluent, CFX, Star CCM+) have shown a remarkable potential for predicting the flow behavior and performance of Savonius-style wind turbines. The computational methodology, more explicitly the selections of turbulence model, grid size, and boundary layer formed on the turbine blades are regarded as the most important criteria. Among the various k-ε and k-ω turbulence models, an effective approach is to solve the unsteady flow governing equations along with the SST k-ω turbulence equations (suggested by Menter [47]) to determine the unknown pressure and velocity terms at various locations of the flow field including the blade geometry, which are used to calculate moment, drag, lift, turbulence, and so forth around the turbine. The detailed computational approaches are discussed in Chapter 6.

4.6 CONCLUDING REMARKS AND RECOMMENDED SAVONIUS DESIGN

Arguing the options for small, low-cost, easily operated, and easily maintained wind energy conversion systems for rural and urban uses, the modified Savonius-style wind turbines are among the solutions with performance efficiencies of ~30%. It can be said that out of all the renewable machines, the Savonius-style turbine is the simplest and of lowest cost but it has comparatively low efficiency and a low power output. This class of turbines has a high solidity factor compared to other well-established counterparts of lift-type convertors, and thus, is not suitable for large-scale power generation. However, with a lighter blade material, an effective system of Savonius-style turbines may be designed for small-scale applications where the advantages of confined space of installation, low wind speed, low cost and simple construction, decentralized power generation, and the direction-independent operation of turbines are the prime concerns.

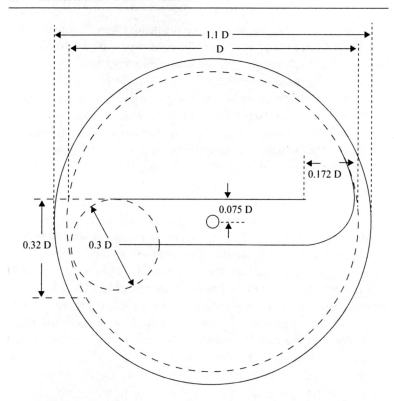

Figure 4.22. Design and dimensions of a Savonius-style wind turbine.
Source: Reproduced from Benesh [28].

With a recognition of the large number of design parameters that impact Savonius performance and the wide range over which those parameters vary, it is helpful to provide a recommended design using the best knowledge of the art. That design, shown in Figure 4.22, includes size, shape, overlap, offset, and plate sizes.

REFERENCES

[1]. Savonius, S. J. (1931). The S-rotor and its applications, *Mechanical Engineering*, 53(5), 333–338.
[2]. Abraham, J.P., Plourde, B.D., Mowry, G.S., Minkowycz, W.J., & Sparrow, E.M. (2012). Summary of Savonius wind turbine development and future applications for small-scale power generation, *Journal of Renewable and Sustainable Energy*, 4(4), 042703, 1–21.

[3]. Akwa, J.V., Vielmo, H.A., & Petry, A.P. (2012). A review on the performance of Savonius wind turbines, *Renewable and Sustainable Energy Reviews*, 16(5), 3054–3064.

[4]. Roy, S., & Saha, U. K. (2013a). Review of experimental investigations into the design, performance and optimization of the Savonius rotor, *Proceedings of IMechE Part A: Journal of Power and Energy*, 227(4), 528–542.

[5]. Shepherd, W., & Zhang, L. (2011). *Electricity Generation Using Wind Power*. Singapore: World Scientific Publishing Co. Pte. Ltd.

[6]. Abraham, J.P., Plourde, B.D., Mowry, G.S., Sparrow, E.M., & Minkowycz, W.J. (2011). Numerical simulation of fluid flow around a vertical-axis turbine, *Journal of Renewable and Sustainable Energy*, 3, 033109, 1–13.

[7]. Plourde, B. D., Abraham, J. P., Mowry, G.S., & Minkowycz, W. J. (2012). Simulations of three-dimensional vertical-axis turbines for communications applications, *Wind Engineering*, 36, 443–454.

[8]. Plourde, B. D., Abraham, J. P., Mowry, G. S., & Minkowycz, W. J. (2011). Use of small-scale wind energy to power cellular communication equipment, *Sensors & Transducers*, 13, 53–61. http://www.sensorsportal.com/HTML/DIGEST/P_SI_167.htm.

[9]. Roy, S., & Saha, U.K. (2013b). Review on the numerical investigations into the design and development of Savonius wind rotors, *Renewable and Sustainable Energy Reviews*, 24, 73–83.

[10]. Alexander, A.J., & Holownia, B.P. (1978). Wind tunnel test on a Savonius rotor, *Journal of Wind Engineering and Industrial Aerodynamics*, 3(4), 343–351.

[11]. Baird, J. P., & Pender, S. F. (1980). Optimization of a vertical axis wind turbine for small scale applications. *Proceedings of the 7th Australasian Hydraulic and Fluid Mechanics Conference*, Brisbane, Australia, August 18–22, 1980.

[12]. Bergeles, G., & Athanassiadis, N. (1982). On the flow field of the Savonius rotor, *Journal of Wind Engineering*, 6(3), 140–148.

[13]. Sivasegaram, S., & Sivapalan, S. (1983). Augmentation of power in slow-running vertical-axis wind rotors using multiple vanes, *Wind Engineering*, 7(1), 12–19.

[14]. Bowden, G. J., & McAleese, S. A. (1984). The properties of isolated and coupled Savonius rotors, *Wind Engineering*, 8(4), 271–288.

[15]. Ogawa, T., & Yoshida, H. (1986). Effects of a deflecting plate and rotor end plates on performance of Savonius type wind turbine, *Bulletin of JSME*, 29(253), 2115–2121.

[16]. Fujisawa, N., & Gotoh, F. (1992). Pressure measurements and flow visualization study of a Savonius rotor, *Journal of Wind Engineering and Industrial Aerodynamics*, 39(1–3), 51–60.

[17]. Huda, M.D., Selim, M.A., Islam, A.K.M.S., & Islam, M.Q. (1992). The performance of an S-shaped Savonius rotor with a deflecting plate, *RERIC International Energy Journal*, 14(1), 25–32.

[18]. Hayashi, T., Li, Y., Hara, Y., & Suzuki, K. (2005). Wind tunnel tests on a three-stage out-phase Savonius rotor, *JSME International Journal Series B:*

Special Issue on Experimental Mechanics in Heat and Fluid Flow, 48(1), 9–16.

[19]. Kamoji, M. A., Kedare, S. B., & Prabhu, S. V. (2008). Experiments investigations on single stage, two stage and three stage conventional Savonius rotor, *International Journal of Energy Research*, 32(10), 877–895.

[20]. Dobrev, I., & Massouh, F. (2011). CFD and PIV investigation of unsteady flow through Savonius wind turbine, *Energy Procedia*, 6, 711–720.

[21]. Gasch, R. & Twele, J. (2012). Scaling wind turbines and rules of similarity. In Gasch, R. & Twele, J. (Ed.), *Wind Power Plants: Fundamentals, Design, Construction and Operation*. Springer Berlin Heidelberg, 257–271.

[22]. Mojola, O.O. (1985). On the aerodynamic design of the Savonius wind mill rotor, *Journal of Wind Engineering and Industrial Aerodynamics*, 21(2), 223–231.

[23]. Saha, U.K., Thotla, S., & Maity, D. (2008). Optimum design configuration of Savonius rotor through wind tunnel experiments, *Journal of Wind Engineering and Industrial Aerodynamics*, 96(8–9), 1359–1375.

[24]. Blackwell, B.F., Sheldahl, R.E., & Feltz, L.V. (1977). Wind tunnel performance data for two- and three-bucket Savonius rotors, Sandia Laboratories, Sand 76-0131 under act AT/29-11.

[25]. Savonius, S.J. (1929). Rotor adapted to be driven by wind or flowing water. US Patent No. 1,697,574.

[26]. Savonius, S.J. (1930). Wind rotor. US Patent No. 1,766,765.

[27]. Benesh, A.H. (1988). Wind turbine system using a vertical axis Savonius-type rotor. US Patent No. 4,784,568.

[28]. Benesh, A.H. (1996). Wind turbine with Savonius-type rotor. US Patent No. 5,494,407.

[29]. Rahai, H.R., & Hefazi, H. (2008). Vertical axis wind turbine with optimized blade profile. US Patent No. 7,393,177 B2.

[30]. Mohamed, M.H., Janiga, G., Pap, E., & Thevenin, D. (2011). Optimal blade shape of a modified Savonius turbine using an obstacle shielding the returning blade, *Energy Conversion and Management*, 52(1), 236–242.

[31]. Plourde, B.D., Abraham, J.P., Mowry, G.S., & Minkowycz, W.J. (2011). An experimental investigation of a large, vertical-axis wind turbine: effects of venting and capping, *Wind Engineering*, 35, 213–220.

[32]. Grinspan, A.S., Saha, U.K., & Mahanta, P. (2004). Experimental investigation of twisted bladed Savonius wind turbine rotor, *International Energy Journal*, 5(1), 1–9.

[33]. Saha, U.K., & Rajkumar, M.J. (2006). On the performance analysis of Savonius rotor with twisted blades, *Renewable Energy*, 31(11), 1776–1788.

[34]. Kamoji, M.A., Kedare, S.B., & Prabhu, S.V. (2009). Performance tests on helical Savonius rotor, *Renewable Energy*, 34(3), 521–529.

[35]. Crissy Field Small VAWT Demonstration (2012). Retrieved November 1, 2013, from http://www.wind-works.org/cms/index.php?id=510.

[36]. SolucionesEnergéticas S. A. (2013). Helical 3 kW wind generator. Retrieved November 1, 2013, from http://www.solener.com/novedad_e.html.
[37]. Grinspan, A.S. (2002). Development of a low speed wind tunnel and testing of Savonius wind turbine rotor with twisted blades. M. Tech Thesis, Department of Mechanical Engineering, IIT Guwahati, India.
[38]. Damak, A., Driss, Z., & Abid, M. S. (2013). Experimental investigation of helical Savonius rotor with a twist of 180°. *Renewable Energy*, 52, 136–142.
[39]. Modi, V.J., & Fernando, M.S. U.K. (1989). On the performance of the Savonius wind turbine, *ASME Journal of Solar Energy Engineering*, 111(1), 71–81.
[40]. Islam, M., Ting, D.S.K., & Fartaj, A. (2008). Aerodynamic models for Darrieus-type straight-bladed vertical axis wind turbines, *Renewable and Sustainable Energy Reviews*, 12(4), 1087–1109.
[41]. Paraschivoiu, I. (2002). *Wind turbine design: with emphasis on Darrieus concept*. Canada: Presses inter Polytechnique.
[42]. Biadgo, A. M., Simonovic, A., Komarov, D., & Stupar, S. (2013). Numerical and analytical investigation of vertical axis wind turbine. *FME Transactions* 41, 49–58.
[43]. Wilson, R.E., Lissaman, P.B.S., & Walker, S.N. (1976). Aerodynamic performance of wind turbines, ERDA/NSF/04014-7611, 111–164.
[44]. Van Dusen, E.S., & Kirchhoff, R.H. (1978). A two dimensional vortex sheet model of a Savonius rotor. Fluid Engineering in Advanced Energy Systems. In *Proceedings of the ASME Winter Annual Meeting*, San Francisco, California, December 10–15, 1978.
[45]. Ogawa, T. (1984). Theoretical study on the flow about Savonius rotor, *ASME Transactions: Journal of Fluids Engineering*, 106(1), 85–91.
[46]. Kotb, M.A., & Aldoss T.K. (1991). Flow field around a partially-blocked Savonius rotor, *Applied Energy*, 38(2), 117–132.
[47]. Menter, F. R. (1994). Two-equation eddy-viscosity turbulence models for engineering applications, *AIAA Journal*, 32(8), 1598–1605.

CHAPTER 5

DESIGN OF HORIZONTAL-AXIS WIND TURBINES

M. Refan, H. Hangan

Here, the authors present an up-to-date analysis of the principal components of horizontal-axis wind turbines (HAWTs). They discuss how these components are designed and how their performance impacts the overall system. Then, the authors describe the performance characteristics of HAWTs and what circumstances make their employment viable.

Various design methodologies are described in detail including the Blade Element Momentum (BEM) theory. Corrections to relate theoretical predictions and practical applications are provided. Finally, issues related to testing methodology are discussed and best practices are given.

The reader will become well versed in all aspects of design, optimization, application, and testing of small-scale horizontal wind turbines.

5.1 SMALL HORIZONTAL-AXIS WIND TURBINES

HAWTs, with their rotation axis parallel to the ground, are the most common design for wind turbines in wind energy technology today. Figure 5.1 shows the performance data for rotors of various configurations. The most obvious advantage of modern rotors, as compared with traditional rotors, is the maximum power coefficient (C_p) achievable. The performance of a wind turbine is generally characterized by the power coefficient, which is the ratio between the rotor power output and the power available in the wind. Modern rotors (Darrieus and HAWTs) are lift-driven machines while the historical windmills (Savonius, American wind turbine, and Dutch windmill) are drag driven. Figure 5.1 clearly demonstrates the advantage of using aerodynamic lift over drag to generate power: Modern lift-driven rotors can achieve a power coefficient of about 0.5 which is

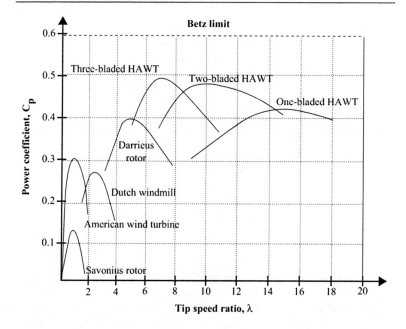

Figure 5.1. Rotor power coefficient as a function of tip speed ratio for various rotor configurations.

Source: Adapted from Wilson et al. [1].

67% more than that of the historical windmills. On the other hand, fast few-bladed rotors (Darrieus and HAWTs) have a very low torque, particularly at the start.

The use of small HAWTs in rural areas is growing and there have been a number of recent developments in small turbines that make them suitable for home use. Although large wind turbines can provide reliable, cost-effective, and competitive power, and their technology is almost mature, they often raise social issues related to their implementation near inhabited rural or residential areas. Small wind turbines, on the other hand, are better accepted in these areas as their presence is almost inconspicuous. However, the performance of these small wind turbines is much less clear compared to their large counterparts. Sahin et al. [2] showed that for a region with annual average wind speed of about 7 m/s (16 mph), the annual energy generation of a 100 W HAWT will be about 150 kW/h, which is a considerable amount of energy for a small, 60-cm- (24 inches)-diameter rotor.

There is no universal standard to classify the small HAWTs. According to the National Renewable Energy Laboratory (NREL), wind turbines with rated power output up to 100 kW and rotor diameter up to 19 m (62 ft) are categorized as small wind turbines [3]. Clausen and Wood [4] defined a small wind turbine as having a maximum power output of 50 kW while Refan and Hangan [5] considered rotors with rated power output from 300 W to 300 kW as small wind turbines. As mentioned in Chapter 1, small wind power referred in this book is in the few kW range.

Despite large HAWTs that are located in areas selected based on optimum wind conditions, small wind turbines are situated in areas that do not necessarily have the best wind conditions. Therefore, the designers should primarily concentrate on achieving good performance characteristics with their rotor design. Although small HAWTs are often referred to as smaller versions of the propeller-style large grid-connected turbines, the airfoil design/selection for these wind turbines is significantly different than for the large wind machines. The performance degradation and noise from laminar separation bubbles are some of the main concerns in the design process of small HAWTS. Therefore, the most efficient airfoil design for larger machines cannot be simply transferred to the small machines.

5.1.1 COMPONENTS AND CURRENT DESIGNS OF HORIZONTAL-AXIS WIND TURBINES

A small HAWT consists of various components as shown in Figure 5.2. A short description of principal subsystems follows.

5.1.1.1 Rotor

The rotor includes blades and the hub, which are the most important components from the performance and cost points of view. Currently, most wind turbines are three bladed with an upwind rotor configuration. Furthermore, fixed blade pitch and stall control are widely used for small HAWTs and the blades are mostly made from fiberglass or carbon fiber.

5.1.1.2 Nacelle and Yaw System

The nacelle is a casing that holds shafts, a mechanical brake, and the generator. The majority of wind turbines use either induction or synchronous generators. Free yaw systems are commonly used in small HAWTs. These systems utilize the wind force to adjust the orientation of the wind turbine rotor into the wind. A simple free yaw system comprises a roller bearing

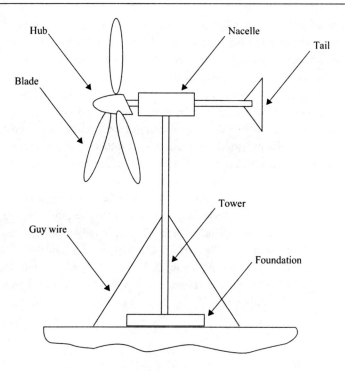

Figure 5.2. Components of a small HAWT.

connection between the tower and the nacelle and a tail fin mounted on the nacelle. The tail turns the wind turbine rotor into the wind by exerting a corrective torque to the nacelle.

5.1.1.3 Tower and Foundation

The tower height is typically at least 1 to 1.5 times the rotor diameter and is attached to a supporting foundation. For small wind turbines, narrow pole towers supported by guy wires are widely used. This configuration has the advantage of lower weight, which reduces the cost. The drawback is the limited access around the tower.

A teetering mechanism, installed on the tower, is used to tilt the wind turbine rotor at high wind speeds (see Figure 5.3). The teetering serves both as a power output regulator and a damage protection mechanism. The teetering movement during rotation creates a dynamic damping effect that eliminates most of the system stresses.

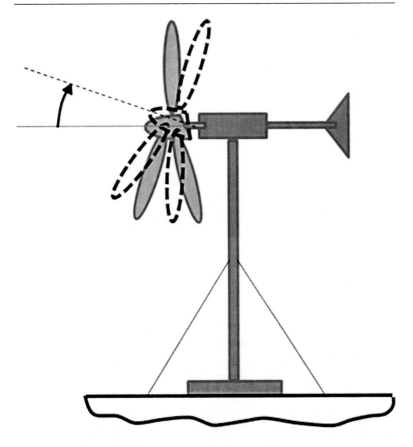

Figure 5.3. Schematic of a teetering mechanism.

5.1.1.4 Control Systems

Small wind turbines utilize a controller to convert the variable three-phase generator output into a direct current (DC) at an appropriate voltage for charging batteries. The machine control system consists of electrical circuits, switches, and flow measurement sensors.

5.1.1.5 Current Designs and Innovations

Figure 5.4 shows the wind turbine currently installed at the rooftop of the Engineering building at the Western University, Canada. A rated power output of 900 W at 13 m/s (29 mph) and a peak power output of 1,200 W

Figure 5.4. A small HAWT installed on the rooftop of the Western University Engineering building.

at 17 m/s (38 mph) are reported by the manufacturer for this rotor. This turbine has the most typical design and configuration for small HAWTs. Turbines with similar design and configuration can be installed at the rooftop of a building or on the ground. However, they should be positioned away from trees or other objects.

A significant variety of design is observed for small HAWTs. Integrating wind turbines in residential and commercial buildings, reducing the cost to the lowest per kWh installed, as well as increasing the efficiency of the rotor is the motivation for recent innovative designs. Some of these conceptual designs are briefly explained.

Figure 5.5 shows seven separate rotors, 2.1 m (6.9 ft) in diameter each, installed coaxially at regular intervals on a single shaft. This way, the swept area and as a result the power production of the wind turbine increases without increasing the diameter of the rotor. This configuration produces 4,500 W in winds of 12 m/s (27 mph) compared to 690 W for one rotor [6]. These turbines were built and tested in California in part funded by the California Energy Commission. The main challenge for this design is to space and angle rotors in a way that each one catches fresh wind.

AeroVironment's small, modular wind turbine, called Architectural Wind, provides an attractive clean energy source that integrates easily into commercial and industrial buildings. Architectural Wind takes advantage of the natural wind acceleration resulting from the aerodynamic characteristics of the building. When wind hits a building, a separation region forms on the rooftop. This results in an accelerated airflow straight up the side of

Figure 5.5. Sky Serpent design installed in Tehachapi, California [8].

the building. This accelerated wind can increase the electrical power generation of the turbine by more than 50% compared to the power generation that would result from systems situated outside of the acceleration zone [7].

A variety of buildings have installed rows of these turbines, including the Maui Ocean Center in Hawaii and Logan International Airport in Boston. An array of 20 wind turbines, each 1.8 m (5.9 ft) tall, was placed at the edge of a rooftop at Logan International Airport in Boston. These turbines are expected to generate about 100,000 kWh per year, equivalent to 3% of the building energy needs [9].

Honeywell wind turbine is a rooftop wind turbine that works at wind speeds as low as 3 km/h (1.9 mph). This gearless turbine creates power from magnets in its blade tips and in the enclosure for the blades. The WT6500 model is able to produce 2752 kWh/y [10]. This is more than 20% of average household annual electricity needs. This model can be installed on a separate pole, tower, or roof. Due to its configuration, it does not need to be positioned away from trees or other objects.

5.1.1.6 Power Output Prediction

The energy production of a wind turbine is a function of the wind speed and is presented as a power performance curve. This curve is unique for each wind turbine design and provides the opportunity to predict the power output of a wind turbine at various wind speeds.

Figure 5.6 displays an example of a power curve for a wind turbine as a function of the wind speed at the hub height. The cut-in speed is the minimum wind speed at which the turbine produces useful power. Another key velocity is the rated wind speed at which the maximum power output of the generator is achieved. The maximum wind speed at which the machine is allowed to deliver power is called the cut-out speed. This speed is limited by safety concerns and design characteristics.

The power curve of a wind turbine is derived from field or wind tunnel measurements and is usually provided by the manufacturer. It is possible to estimate the power curve of a turbine through calculations that require information about the mechanical and electrical efficiencies of the components as well as aerodynamic characteristics of the blades.

5.1.2 DESIGN, TESTING, AND PERFORMANCE OF SMALL HORIZONTAL-AXIS WIND TURBINES

5.1.2.1 Wind Turbine Design

The wind turbine design process involves combining a great number of electrical and mechanical components while considering the economic

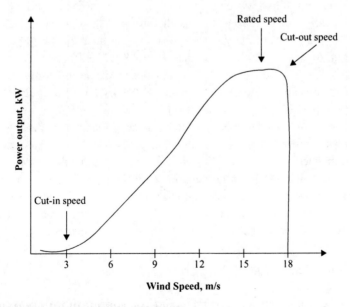

Figure 5.6. Typical power curve of small HAWTs.

viability of the machine. The energy cost of a wind turbine is mainly influenced by the cost of the initial components; the installation, operation, and maintenance costs; and the annual energy productivity. The wind turbine productivity is a function of the design and the wind resources. Therefore, the wind turbine design plays an important role in the overall cost and productivity of the machine.

The design process starts with determining the application. Small HAWTs are mainly aimed for use by utility customers and at remote areas and are typically in the few kW range. For these wind turbines, simplicity of the machine and ease of installation, operation, and maintenance are key factors that need to be considered in the design.

The next step in the design process is to investigate previous attempts to design a wind turbine with similar applications. This will narrow down the options and will avoid unnecessary trial and error.

The overall wind turbine layout or topology is another topic that needs to be addressed. The main choices in the turbine configuration are listed as follows:

- Power control: variable pitch blades or aerodynamic stall
- Hub design: teetering or rigid
- Wind alignment: free yaw or active yaw
- Rotor orientation: upwind or downwind
- Rotor speed: constant or variable
- Number of blades
- Design tip speed ratio and solidity

The summary of these key choices are provided by Manwell et al. [11]. The most common practice for the case of small HAWTs is using an upwind three-bladed rotor with free yaw wind alignment. The speed and power output of a small HAWT are usually controlled by using a stall mechanism because a blade-pitching mechanism is not economically feasible.

After selection of the overall layout of the wind turbine, a tentative design for the rotor, nacelle, tower, and foundation is developed. Each principal component consists of subsystems that need to be addressed in the design process. The rotor performance and its subsystem design process are discussed in detail. The design process for other principal components is explained by Manwell et al. [11] and Hau [12].

Once the preliminary design is developed, it is essential to estimate the power curve of the turbine. The wind turbine performance is affected by the rotor design as well as the generator efficiency and type. In addition, it is important to assess the ability of the wind turbine to withstand various loads such as static, cyclic, transient, and so forth. After evaluating the power curve of the turbine, as well as its ability to withstand encountered loads, the designer may need to revise the design. The final design will be used to build a prototype that will be tested to verify the design concept. Three wind turbine prototype test methods are available: field measurements, wind tunnel testing, and vehicle-based measurements. These methods are reviewed in more detail later in this chapter. The power output is measured, using any of these measurement methods, at various wind speeds and is compared to the estimated power curve. In addition, loads experienced by critical components of the turbine are measured using strain gauges.

5.1.2.2 Theory

There are various mathematical models for performance analysis of the turbine rotor including BEM theory, Vortex Wake method, and acceleration potential. Among these modeling techniques, the BEM theory is the most common mathematical model for evaluating wind turbine rotor performance. There are two reasons for the popularity of this theory: (1) the accuracy of the method for a wide variety of flow conditions and rotors and (2) the relative simplicity of the method [13].

The BEM theory uses the lift and drag coefficients of the turbine blade airfoils to evaluate the performance of the rotor. In recent years, several efforts have been made to modify and optimize the BEM theory to provide accurate and reliable results. Challenges relate to the evaluation of the axial and angular induction factors (a and a', respectively), curve fitting of the lift and drag coefficients, as well as benchmarking against experimental data to assess and improve the theory. This lack of benchmarking is a common challenge with all wind turbine performance evaluation techniques [14].

The BEM theory has proved to be accurate for a wide range of large rotors and flow conditions and is the most popular model for preliminary assessment of a wind turbine rotor performance. This theory is two-dimensional, which means that in the case of small HAWTs, where the three-dimensional effects are more significant (compared to large HAWTs) due to smaller blade span, the BEM theory may not work well. Therefore, evaluation of the BEM theory performance in predicting aerodynamic behavior of a small HAWT rotor was essential. Until very recently, no research was found to address the performance of the BEM theory in the case of small wind turbines. Refan and Hangan [5] investigated the accuracy of the BEM theory through wind tunnel testing of a three-bladed small HAWT. The comparison showed that the overall prediction of the theory is within acceptable range of accuracy. However, the BEM theory prediction for the small wind turbine was not as accurate as the prediction for larger wind turbines, and the assumption of negligible radial flow introduced significant errors in the theoretical analysis especially for low- and mid-range wind speeds.

In the following paragraphs, the theory related to aerodynamic design of a small HAWT is detailed. This is intended for those readers interested in the detailed design concepts. The following section deals with the general characteristics and design of the rotor for small HAWTs.

The aerodynamic behavior analysis of wind turbines starts without any specific design. The general device that is used to analyze the wind turbine performance based on the energy extraction process is called an *actuator disc* (see Figure 5.7).

5.1.2.2.1 Linear Momentum

Betz [15] introduced a simple model, adapted from the linear momentum theory that estimates ship propeller performance, to determine the power and the torque of an ideal wind turbine. In this model, the rotor is replaced by an actuator disc that creates discontinuity of pressure. Furthermore, the wake rotation effects are neglected. In this analysis, it is assumed that

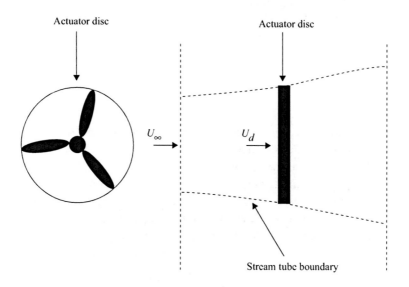

Figure 5.7. Wind turbine rotor modeled by an actuator disc.

frictional drag is zero and the thrust on the disk area is uniform. For the full list of the assumptions used in this analysis, see Manwell [11].

The conservation of linear momentum is applied to a control volume in which the boundaries are the surface and two cross-sections at the ends of a stream tube. The resulting force on the control volume is equal and opposite to the net force applied by wind on the rotor and is called thrust (T).

The axial thrust of the wind on the ideal rotor (disk) can be expressed as a function of air density (ρ), upstream wind velocity (U_∞), cross-sectional area (A), and axial induction factor (a).

$$T = \frac{1}{2}\rho A U_\infty^2 4a(1-a) \tag{5.1}$$

Where a is defined as the ratio of the decrease in wind velocity between the upstream and the disk plane (U_d) to the free stream velocity.

$$a = \frac{U_\infty - U_d}{U_\infty} \tag{5.2}$$

Therefore, the rotor power output can be calculated by multiplying the thrust with the wind speed at the rotor section.

$$P = \frac{1}{2}\rho A U_\infty^3 4a(1-a)^2 \tag{5.3}$$

The nondimensional parameter that characterizes the rotor performance is called the power coefficient (C_p). It is defined as the ratio between the rotor power output and the power available in the wind.

$$C_P = \frac{P}{\frac{1}{2}\rho A U_\infty^3} = 4a(1-a)^2 \tag{5.4}$$

The maximum power coefficient that is theoretically achievable is called Betz limit and is equal to 0.5926. Note that this limit is not caused by any deficiency in design, as the model is not representative of any particular type of a wind turbine. The maximum possible power coefficient decreases in practice as a result of (1) aerodynamic drag, (2) wake rotation, (3) finite number of blades, and (4) blade tip losses.

The overall efficiency of the turbine is a function of mechanical and electrical efficiencies of the wind machine and the rotor power coefficient:

$$\eta_0 = \eta_{mech,elec} C_P \tag{5.5}$$

The primary assumption of the previous analysis was a nonrotating wake. When the wind turbine rotates, the flow in the wake rotates as well but in an opposite direction. This is due to the torque applied by the flow on the rotor. The rotational wake reduces the energy extracted by the rotor. Furthermore, the air particles in the wake of the rotor gain a velocity tangential to the plane of rotation. This extra velocity component increases the kinetic energy of the wake, which results in a fall in the static pressure of the air and reduces the maximum extractable energy by the rotor. Consequently, the power coefficient is less than Betz's limit. In addition, the power coefficient then becomes a function of the ratio between axial and angular motions of the air flow, which is called the tip speed ratio (λ).

The wind can pass undisturbed through the gaps between the blades while the rotor is rotating slowly. On the other hand, a rapidly rotating rotor looks like a solid disk to the flow. Therefore, to obtain optimal rotor efficiency, it is essential to relate the angular velocity of the rotor (Ω) to the free stream wind speed. The relationship between these two speeds is characterized by the tip speed ratio.

$$\lambda = \frac{R\Omega}{U_\infty} \qquad (5.6)$$

where R is the rotor radius. By relating the time taken for a blade to move into its predecessor location to the time taken for the perturbed wind to restore, the optimal tip speed ratio (λ_{opt}) for maximum power extraction can be found. This optimal speed ratio is a function of the number of the blades in a rotor. The fewer the number of blades, the higher is the optimal tip speed ratio as the rotor has to rotate faster to extract the maximum power from the wind. The optimal tip speed ratios for various numbers of blades are listed in Table 5.1.

Table 5.1. The optimal tip speed ratio for various blade configurations

Number of Blades, B	Optimal Tip Speed Ratio, λ_{opt}
> 12	1
6–12	2
4–5	3
3	4
2	6

5.1.2.2.2 Angular Momentum

Turbine rotation results in a variable tangential velocity along the blade span. Therefore, the stream tube model needs to be modified to an annular ring of the rotor disc, which is of radius r and of radial width dr (see Figure 5.8).

The flow just upstream of the actuator disc is free of rotation. If the angular velocity of the rotor is considered to be Ω, then the flow exiting the disc has a tangential velocity equal to $2r\Omega a'$, where a' is the angular induction factor representing the tangential velocity changes in the flow. Figure 5.9 shows characteristics of the flow while passing through blades.

The driving torque (Q) on the rotor shaft is the same as the torque on the ring and will be equal to the rate of change of angular momentum of the air flow passing through the rotor (conservation of angular momentum). Therefore, on a small annular area element dA, the torque is

$$dQ = \rho U_d dA (2a' r\Omega) r \tag{5.7}$$

and the rotor shaft power output is

$$dP = dQ(\Omega) \tag{5.8}$$

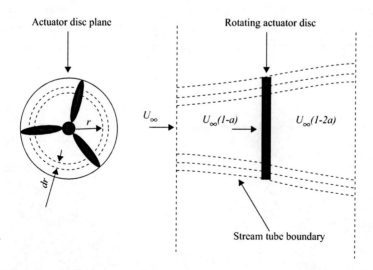

Figure 5.8. Schematic of the rotating actuator disc with an annular ring of radius r.

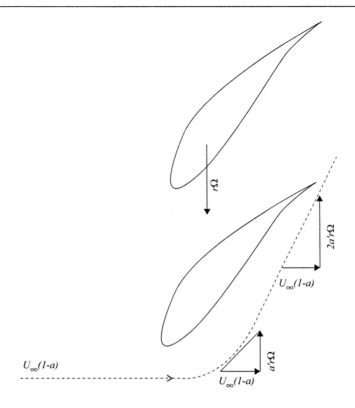

Figure 5.9. Tangential velocity changes across the rotor disc.
Source: Adapted from Burton et al. [16].

Note that in these analyses it is assumed that each ring acts independently in imparting momentum into the air.

Combining Equations (5.7) and (5.8) and using the definition of local speed ratio (λ_r), the power at each annular ring can be expressed as in Ref. [16].

$$dP = 4\pi\rho U_\infty^3 a'(1-a)\lambda_r^2 r dr \quad (5.9)$$

Therefore, the power at each annular ring is a function of the axial and angular induction factors and the tip speed ratio.

The overall power coefficient can be determined after calculating the incremental contribution of each annular ring to the power coefficient and integrating over the blade span

$$C_P = \frac{8}{R^2} \int_{r=0}^{r=R} a'(1-a)\lambda_r^2 r dr \quad (5.10)$$

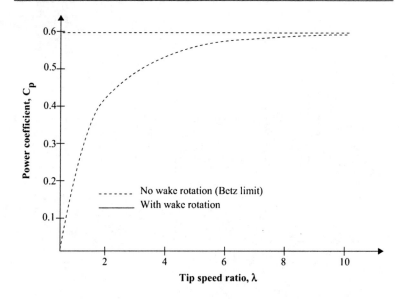

Figure 5.10. Theoretical maximum power coefficient as a function of tip speed ratio for an ideal HAWT with and without wake rotation

Figure 5.10 shows the theoretical maximum power coefficient of an ideal HAWT with and without wake rotation. It is seen that by increasing the tip speed ratio, the maximum theoretical C_p increases.

5.1.2.2.3 Blade Element Theory

This theory determines the steady state aerodynamic loading and the rotor power output for a given blade geometry by integrating the thrust distribution and the tangential force distribution over the rotor in the plane of rotation, respectively.

The aerodynamic lift and drag forces on a blade section of radius r and radial width dr are accountable for the force experienced by the same blade section due to the rotational wake as well as for the rate of change of momentum of the air that passes the swept annulus. Furthermore, the rotational velocity in the wake results in a pressure drop that causes a force on the blade element which must also be provided by aerodynamic lift and drag.

It is assumed that the forces on a blade can be expressed as a function of two-dimensional airfoil characteristics. For this purpose, the blade is divided into N sections and the following assumptions are made:

- The velocity component in the span-wise direction is ignored
- The radial aerodynamic interaction between elements is ignored
- Three-dimensional effects are ignored

The effective or resultant wind is a vector sum of the wind velocity at the rotor plane and the net tangential velocity of the blade. The net tangential velocity of the blade section is a combination of the tangential velocity of the blade element ($r\Omega$) shown in Figure 5.11 and the tangential velocity of the wake ($r\Omega a'$). Thus, the resultant relative velocity (U_{rel}) at each blade element is

$$U_{rel}^2 = \left[U_x(1-a)\right]^2 + \left[r\Omega(1+a')\right]^2 \qquad (5.11)$$

Figure 5.12 shows the blade geometry for the current analysis. θ is the twist angle of the blade section, α is the angle of attack (the angle between chord line and the resultant relative wind), and φ is the angle of

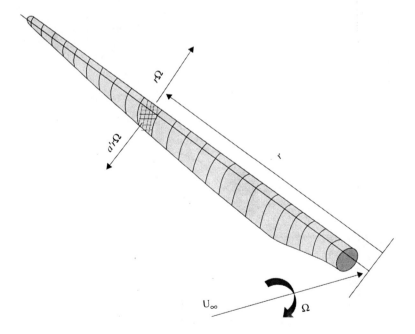

Figure 5.11. Blade geometry for blade element analysis.

the resultant relative wind. Furthermore, dF_L and dF_D are the incremental lift and drag forces, respectively. dF_N is the incremental force normal to the plane of rotation that contributes to thrust and dF_T is the incremental force tangential to the plane of rotation that creates useful torque.

The following relationships can be determined from Figure 5.12.

$$\tan\varphi = \frac{U_\infty(1-a)}{r\Omega(1+a')} \tag{5.12}$$

$$U_{rel} = \frac{U_\infty(1-a)}{\sin\varphi} \tag{5.13}$$

$$dF_N = dF_L \cos\varphi + dF_D \sin\varphi \tag{5.14}$$

$$dF_T = dF_L \sin\varphi - dF_D \cos\varphi \tag{5.15}$$

Also, it can be shown that

$$C_l = \frac{dF_L}{\frac{1}{2}\rho U_{rel}^2 cdr} \tag{5.16}$$

$$C_d = \frac{dF_D}{\frac{1}{2}\rho U_{rel}^2 cdr} \tag{5.17}$$

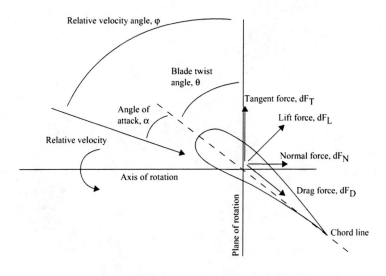

Figure 5.12. Blade element geometry, velocities, and forces [5].

Where C_l and C_d are lift and drag coefficients, respectively, and c is the chord length.

If the rotor has B blades, then the total differential torque due to the tangential force on the section at a distance r from the center is

$$dQ = BrdF_T = B\frac{1}{2}\rho U_{rel}^2 \left(C_l \sin\varphi - C_d \cos\varphi\right) crdr \qquad (5.18)$$

and the total normal force at the same section is

$$dF_N = B\frac{1}{2}\rho U_{rel}^2 \left(C_l \cos\varphi + C_d \sin\varphi\right) cdr \qquad (5.19)$$

Hence, the blade element theory results in two equations expressing the normal and the tangential force on the annular rotor sections. These equations along with other equations and with additional assumptions are used to find the rotor performance of a HAWT.

5.1.2.2.4 Blade Element Momentum Theory

The BEM theory assumes that the change in the momentum of the air from upstream to downstream of the annulus swept by the element is due to the forces of the blade. Thus, the forces and moments derived from axial and angular momentum theories and blade element theory must be equal. For this analysis, twist and chord distributions of the blade are known parameters while the unknown is the angle of attack and performance of the rotor, which can be solved using BEM results and additional relationships.

By applying the conservation of linear and angular momentum to an annular element of the rotor disk (which is of radius r and of radial width dr), the thrust (T) and torque (Q) on the annular ring of the rotor can be found as a function of axial and angular induction factors, a and a'.

$$dT = \rho U_x^2 4a(1-a)\pi r\, dr \qquad (5.20)$$

$$dQ = \rho U_x 4a'(1-a)\pi r^3 \Omega\, dr \qquad (5.21)$$

Furthermore, by applying the blade element theory for a blade element at radius r and introducing the local solidity (σ'), the normal force and the torque are obtained as follows

$$dF_N = \sigma'\pi\rho \frac{U_x^2 (1-a)^2}{\sin^2\varphi}\left(C_l \cos\varphi + C_d \sin\varphi\right) rdr \qquad (5.22)$$

$$dQ = \sigma' \pi \rho \frac{U_\infty^2 (1-a)^2 (C_1 \sin\varphi - C_d \cos\varphi) r^2 dr}{(\sin^2 \varphi)} \quad (5.23)$$

Where σ' is defined as

$$\sigma' = \frac{Bc}{2\pi r} \quad (5.24)$$

By equating normal forces and torques from the blade element theory and momentum theory, the flow parameters, a, a' and α for each blade element can be found.

$$\frac{a}{(1-a)} = \frac{\sigma'}{4\sin^2\varphi} (C_1 \cos\varphi + C_d \sin\varphi) \quad (5.25)$$

$$\frac{a'}{(1-a)} = \frac{\sigma'}{4r\Omega\sin^2\varphi} U_\infty (C_1 \sin\varphi - C_d \cos\varphi) \quad (5.26)$$

Other equations that can be obtained through algebraic manipulation are:

$$a' = \frac{1}{2}\left[\sqrt{1 + \frac{4}{\lambda_r^2} a(1-a)} - 1\right] \quad (5.27)$$

$$\varphi = \tan^{-1}\left(\frac{1-a}{(1+a')\lambda_r}\right) = \alpha + \theta \quad (5.28)$$

The lift and drag coefficients in Equations (5.25) and (5.26) depend on Reynolds number and the angle of attack for each blade section. An iterative solution can be used to determine the flow conditions at each blade element using the BEM equations (Equations 5.25–5.28), which can be used to estimate the power output of a wind turbine.

Figure 5.13 demonstrates the BEM theory performance in estimating the power output of a three-bladed HAWT. It is clear that the theory overestimates the power generation at very low tunnel wind speeds. The hysteresis losses at low wind speeds are high in the generator and the rotor does not start rotating at low wind speeds due to friction in the generator. Therefore, the power generated at low wind speeds is relatively low and the theory overestimates the power output with the maximum difference of 37.5% at 5 m/s (11 mph).

The significant discrepancy observed at around 9 m/s (20 mph) is due to the rotor teetering. This can be easily corrected in the calculations by calculating the change in the area of the rotor perpendicular to the upcoming wind as a result of teetering.

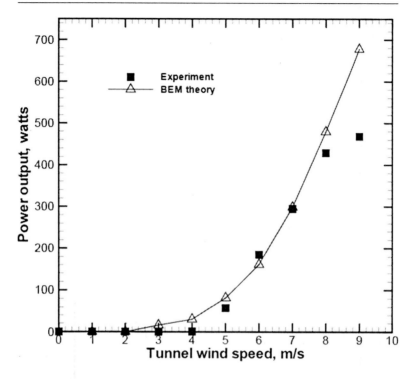

Figure 5.13. BEM prediction of power output compared with wind tunnel measurements [17].

5.1.2.2.5 Tip Loss Correction

The pressure on the lower surface of a blade is less than that on the upper surface. Thus, the air travels around the tip from the lower to the upper surface, which reduces the lift and power generation at the tip region. Prandtl developed a correction factor, F, which is introduced into the previously discussed equations to include the effect of tip losses. This correction factor is shown in Equation (5.29) [11].

$$F = \frac{2}{\pi} \cos^{-1} \left\{ \exp \left[-\left(\frac{\frac{B}{2}(1-\frac{r}{R})}{\frac{r}{R}\sin\varphi} \right) \right] \right\} \qquad (5.29)$$

Therefore, F is a function of the number of blades (B), the angle of relative wind (φ), and the position on the blade (r); it is always between 0 and 1.

This correction factor affects the forces derived from axial and angular momentum theory but does not change the equations from the blade element theory. Thus, the new momentum theory equations are set equal to the previous blade element theory equations, using the BEM theory

$$\frac{a}{(1-a)} = \frac{\sigma'}{4F\sin^2\varphi}(C_1\cos\varphi + C_d\sin\varphi) \quad (5.30)$$

$$\frac{a'}{(1-a)} = \frac{\sigma'}{4Fr\Omega\sin^2\varphi}U_x(C_1\sin\varphi - C_d\cos\varphi) \quad (5.31)$$

So, the power coefficient can be calculated from

$$C_p = \frac{8}{\lambda^2}\int_{\lambda@r=0}^{\lambda@r=R} Fa'(1-a)\left[1-\frac{C_d}{C_1}\cot\varphi\right]\lambda_r^3 d\lambda_r \quad (5.32)$$

Using a sum approximating the integral in Equation (5.32), the power coefficient is determined as follows [16]:

$$C_p = \frac{8}{N\lambda}\sum_{i=1}^{N} F_i a'_i(1-a_i)\left[1-\frac{C_d}{C_1}\cot\varphi_i\right]\lambda_{ri}^3 \quad (5.33)$$

Figure 5.14 compares wind tunnel measurements of the power output of a small three-bladed HAWT with BEM predictions. Applying the Prandtl factor to correct for tip losses reduces the power generated by the rotor, particularly in higher wind speeds.

5.1.2.2.6 Stall Delay

Computational aerodynamic analyses have revealed a decrease in the adverse pressure gradient on a rotating blade. In other words, the angle of attack at which stall occurs is smaller for a stationary blade than a rotating blade. This phenomenon is called stall delay, which was first noticed by Himmelskamp [18] on propellers. The reason for this phenomenon is not clear yet. However, it is agreed that for an unknown reason, the blade rotation reduces the adverse pressure gradient that the flow encounters passing over the downstream edge of the blade.

Ronsten [19] measured blade surface pressures on a stationary and rotating blade and Snel et al. [20] proposed an empirical relationship

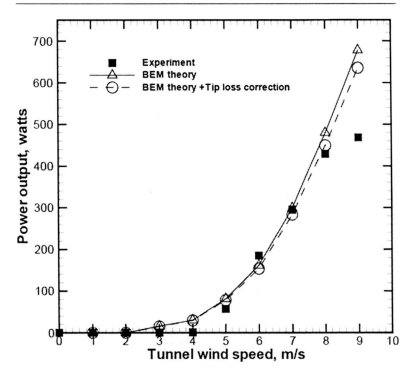

Figure 5.14. Comparison between experimental results and BEM theory prediction with and without tip loss corrections [17].

between two-dimensional and three-dimensional lift coefficients that fit the measurements by Ronsten. The correction equation is as follows [16]:

$$C_{l_{3D}} = C_{l_{2D}} + 3\left(\frac{c}{r}\right)^2 \Delta C_l \qquad (5.34)$$

Where ΔC_l is the difference between two-dimensional C_l-α curve and the linear part of the static curve, which is extended beyond the stall.

Figure 5.15 demonstrates the effect of stall delay corrections on the BEM prediction. It is seen that by increasing the wind speed, the stall delay correction becomes more effective, which means that more sections of the blade are entering the stall development regime. The abrupt change in the trend of the power prediction at 10 m/s (22 mph) is another sign of the change in the regime of the flow. Moreover, the relatively constant slope of the power output reveals that most of the blade sections are still in the attached flow regime and thus, there is no change in the slope of the curve.

5.1.2.2.7 Thrust Coefficient Correction

The measurement of thrust coefficient at various axial induction factors showed that the thrust coefficient calculated for an ideal wind turbine is valid for axial induction factors less than 0.4. Therefore, when the axial induction factor becomes larger than approximately 0.4, the thrust determined by the momentum theory is no longer valid. In this case, the previous analysis can lead to a lack of solution convergence. Different empirical relations between the thrust coefficient and axial induction factor can be adapted to fit the measurements:

An empirical relationship developed by Glauert [21] has the following form:

$$a = \frac{1}{F}\left[0.143 + \sqrt{0.0203 - 0.6427(0.889 - C_T)}\right] \quad (5.35)$$

Where C_T is the thrust coefficient. This equation is valid for $C_T > 0.96$ or, equivalently, for $a > 0.4$.

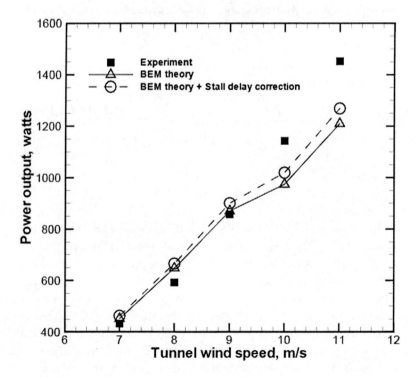

Figure 5.15. Effect of stall delay correction on the rotor performance prediction by BEM [17].

Wilson and Walker [22] also developed an empirical relationship that is valid for $a > 0.2$.

$$a = 0.2[2 + 0.6K - \sqrt{(0.6K + 2)^2 + 4(0.04K - 1)}] \qquad (5.36)$$

Where K is defined as

$$K = \frac{4F\sin^2\varphi}{(F_L \cos\varphi + F_D \sin\varphi)} \qquad (5.37)$$

5.1.2.3 Rotor Design

The BEM theory explained in the previous section can be used to design a wind turbine rotor. The process starts with determining the rotor parameters based on the application and continues with the blade design. Once a proper airfoil is selected, the blade is divided into N elements (usually 10–20) and the chord length is calculated for each element of the blade. The twist, chord, and thickness variations along the span can be considered linear to simplify the fabrication process. The rotor performance characteristics can then be calculated through an iterative solution. The rotor parameters, airfoil selection, and blade shape are explained in more detail in the following sections. However, for a step-by-step instruction of the rotor design process see Manwell et al. [11].

5.1.2.3.1 Rotor Parameters

The following parameters need to be determined by the designer:

1. Rotor radius (R): The power output of a rotor (P) is a function of the power coefficient, efficiencies of various electrical and mechanical systems, wind speed (U_∞), as well as the swept area.

$$P = C_P \frac{1}{2} \rho U_\infty^3 \pi R^2 \qquad (5.38)$$

   By selecting the power that is needed at a certain wind speed and a realistic power coefficient, the rotor radius can be estimated.
2. Tip speed ratio: This parameter depends on the wind turbine application. For applications that require higher torque, $1 < \lambda < 3$. For electric power generation, tip speed ratios up to 10 can be used.
3. Number of blades (B): Table 5.1 can be used to select the number of blades.

5.1.2.3.2 Blade Shape

Selection of airfoils to construct the blade is an important step in the rotor design process. The shape of the airfoil depends on the tip speed ratio. For $\lambda < 3$, curved plates generally work well and for $\lambda > 3$, more aerodynamic airfoils must be used. Furthermore, the airfoil may vary along the blade span: Thick and wide profiles are usually used at the root of the blade and thin profiles with higher aerodynamic performance are used toward the tip.

The blade sections near the hub need to withstand forces and stresses from the rest of the blade. Therefore, thick and wide profiles should be selected for the root region. Furthermore, the blade needs to be tapered along the span toward the tip to counteract an increase in the lift force toward the tip. Overall, the blade profile is a balance between higher aerodynamic characteristics and a better strength. It is seen that [23] for small HAWTs with relatively low Reynolds number, thick airfoil sections (20–25% thickness) have poor performance. On the other hand, thin airfoils with high camber produce more power with lower thrust load [24].

In a comprehensive research study performed by Selig et al. [25] and sponsored by NREL, aerodynamic performances of six airfoils were studied using wind tunnel testing. These airfoils were candidates for use on small wind turbines and are as follows: E387, S822, SD2030, FX-63137, S834, and SH3055. These airfoils represent a broad range of performance characteristics and geometric properties. In this report, the main performance characteristics of each airfoil is described to provide blade designers with information required for decision making. Other airfoils that have been used in small HAWTs include, but not limited to, NACA 6515, NACA 1412, NACA 63-621, FX 66196, S809, SD 7062, and others.

Once the airfoil is selected for each section of the blade, aerodynamic performance data for each airfoil should be obtained. Performance data include lift and drag coefficient variation with the angle of attack (α). Then, the design lift coefficient and the design angle of attack must be selected such that $C_{d,design}/C_{l,design}$ is minimized at each blade section.

Ideally, the aerodynamic characteristics can be determined through wind tunnel testing of an airfoil for a wide range of Reynolds numbers and angles of attack. In a wind tunnel testing setup, the airfoil can also be tested for various surface roughness conditions. The University of Illinois at Urbana-Champaign (UIUC) has a comprehensive database of low-speed airfoils wind tunnel test results [26]. The database includes airfoil geometry coordinates as well as the lift and drag coefficients as a function of the angle of attack at various Reynolds numbers. This database is available to the public free of charge and is an excellent resource for blade designers.

Alternatives to the wind tunnel testing are Computational Fluid Dynamics (CFD) methods and software applications such as XFOIL and JavaFoil, which are interactive programs for the design and analysis of subsonic isolated airfoils. Given the two-dimensional airfoil shape and Reynolds number, these programs can calculate the pressure distribution over the airfoil using the potential flow analysis with a higher-order panel method. Therefore, the lift and drag coefficients can be obtained at any angle of attack and at a certain Reynolds number. Although calculating the aerodynamic performance of the profiles using free software applications is fast and convenient, it is the responsibility of the designer to judge the accuracy of the engineering software. For example, Refan [17] evaluated the accuracy of JavaFoil in calculating lift and drag coefficients of FX 63 137 airfoil by comparing the results with those from wind tunnel experiments. More details on computational methods for evaluating wind turbines are given elsewhere in this book.

Figure 5.16 shows the experimental and calculated lift and drag coefficient variations with the angle of attack. While before stall, the calculated lift curve shows a trend similar to that of the measured one, after

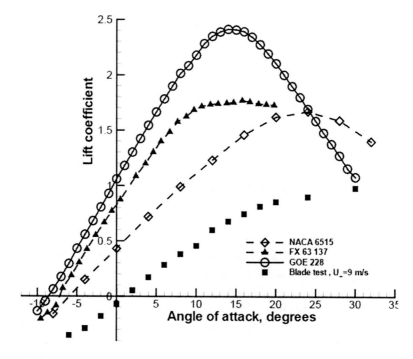

Figure 5.16. Lift and drag coefficients for FX 63 137 airfoil obtained through wind tunnel testing and JavaFoil [17].

stall, significant differences between the calculated and measured lift and drag coefficients are observed. Furthermore, the JavaFoil prediction of the maximum lift coefficient is not accurate. In another research study, Song [27] applied XFOIL to calculate the C_l and C_d of SD 7062 airfoil for $Re = 1 \times 10^5$ to 6×10^5 and compared the results with those provided by UIUC. While the lift coefficients predicted by XFOIL matched the wind tunnel data very well, significant discrepancies were observed in the drag coefficient values.

5.1.2.4 Prototype Testing

Although BEM theory and CFD modeling can be used to predict the power output of a wind turbine, actual measurement of the turbine performance is required to validate the theoretical and numerical results. For the case of small HAWTs, three test methods are available. The most common method is the field measurement. Once the prototype is designed and manufactured, it will be installed in the field and the actual power output of the rotor at various wind conditions will be recorded. Although this is the most realistic way to assess the performance of a rotor, it is very time consuming and location dependent.

Another method to evaluate the performance of a small rotor is called vehicle-based testing [28]. In this method, the wind turbine is installed on a truck and the power output is measured while the truck moves at various wind speeds in a straight course (see Figure 5.17). This way, measurements can be performed in a short period of time and at a very low cost.

As an alternative, a wind turbine prototype can be tested in a wind tunnel. This method has the advantage of controllable conditions and repeatability. The wind tunnel testing is expensive compared to the other methods and is mostly recommended for new design concept evaluations. Furthermore, the main challenge is to have a large wind tunnel that can accommodate a small HAWT with an acceptable blockage ratio (ratio between the wind turbine swept area and the wind tunnel cross-section area). One of the very few full-scale wind tunnel testing studies of small HAWTs was performed at Western University by Refan and Hangan [5]. The study provided insight toward the feasibility of full-scale wind turbine tests in the wind tunnel and its challenges. An upwind, three-bladed small HAWT rotor of 2.2 m (7.2 ft) diameter was selected for the experiments. The rotor was tested in the Boundary Layer Wind Tunnel 2 of the Western University. The wind turbine was first tested in the low-speed test section

Figure 5.17. Schematic drawing of a vehicle-based wind turbine prototype testing setup.

(5 m wide by 3.5 m high, 16.5 by 11.5 ft) of the tunnel and then in the high-speed test section (3.5 m wide by 2.6 m high, 11.5 ft by 8.5 ft) to determine the power output for a wide range of tunnel wind speeds, from 1 m/s to 11 m/s (2.2–24.6 mph).

Figure 5.18 shows the setup in the low-speed section. The wind turbine rotor was installed 11 m (36 ft) downstream of the test section on a 1.7 m (5.6 ft) high pole and the tail was fixed to prevent the yaw motion of the rotor.

The solid blockage caused by the rotor in the low-speed section was about 21% and about 45% in the high-speed section of the tunnel. It is necessary to correct the results as this blockage ratio can cause significant errors in test results. The Glauert correction [29] was selected and applied to determine the "equivalent free airspeed corresponding to the tunnel wind speed, at which the turbine, rotating with the same angular velocity as in the tunnel, would produce the same torque" [30].

The analysis performed by Refan and Hangan [5] revealed that applying blockage corrections leads to reasonable results and, thus, wind tunnel testing of a wind turbine can lead to realistic predictions even for high blockage ratios.

Figure 5.18. Wind tunnel test setup of a small HAWT [17].

5.1.2.5 Performance

The performance of a wind turbine can be characterized by three nondimensional characteristic curves; power, torque, and thrust coefficients as a function of tip speed ratio.

The power coefficient is defined as the rotor power divided by the power available in the wind. Therefore, it determines the amount of power captured by the rotor. As mentioned before, the maximum achievable power coefficient for a wind turbine is 0.5926, the Betz limit. However, the actual power coefficient is always less than the Betz limit due to tip losses, wake rotation effects, and drag. The power coefficient of a typical upwind three-bladed rotor is compared with the Betz limit in Figure 5.19. It is seen that the power coefficient varies with the tip speed ratio. The maximum power extraction is achieved at the optimal tip speed ratio where the difference between the Betz limit and the actual C_p curve is minimal. The hashed area in Figure 5.19 shows the wind power that is not captured with the rotor.

The torque coefficient (C_Q) is calculated by dividing the power coefficient by the tip speed ratio. This nondimensional parameter does not provide any additional information about the performance of the wind turbine. In the case of large HAWTs, the C_Q–λ curve is used to determine the torque on the shaft and the size of the gearbox.

The thrust coefficient (C_T) represents the rotor thrust that is important in the structural design of the tower. The thrust force on the turbine rotor is applied to the tower and then to the foundation/supports of the turbine. Therefore, it plays an important role in the structural design process.

In general, solidity (σ) is a principal parameter that affects the performance of wind turbines. Low solidity results in a smaller maximum power coefficient that is relatively constant over a wide range of tip speed ratios. On the other hand, high solidity produces a sharp peak at a higher C_p value. Therefore, the rotor performance is highly sensitive to the tip speed ratio. Two- and three-bladed rotors have been shown to have the optimum solidity. In addition, the thrust experienced by the rotor and the torque developed increase as the solidity of the rotor increases.

Figure 5.20 compares the experimental (wind tunnel measurements) and theoretical power coefficients as a function of tip speed ratio for the small HAWT tested at Western University. It is seen that the rotor operates at tip speed ratios between 8 and 11. This range of tip speed ratio is significantly higher than the optimal tip speed ratio of a three-bladed wind turbine, which is $\lambda = 4.2$.

Figure 5.21 shows the experimental and theoretical power curve of the previously mentioned wind turbine as a function of the wind speed, which is corrected for the tunnel blockage, tip losses, and stall delay

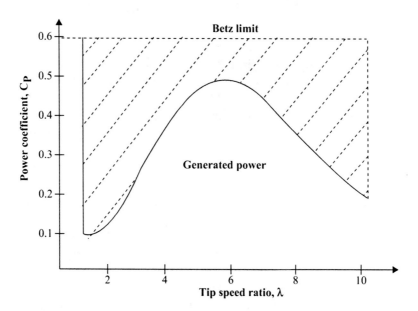

Figure 5.19. Power coefficient variation with tip speed ratio for a three-bladed HAWT.

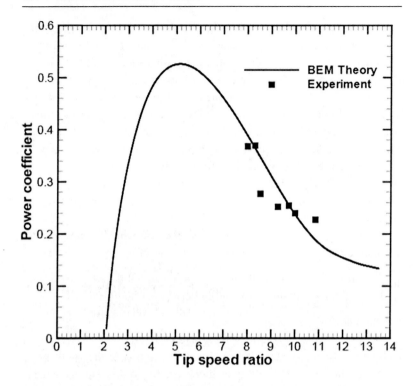

Figure 5.20. Comparison of experimental and theoretical power coefficients [17].

phenomenon. Furthermore, the percentage of the difference between the power predicted (by BEM theory) and the one measured (in the wind tunnel) is depicted in Figure 5.22 for different wind speeds.

Figure 5.22 shows that the maximum discrepancy between theory and measurements is 37.5%, at $U_\infty = 5$ m/s (11.2 mph) and the minimum discrepancy is 3.9% at $U_\infty = 7$ m/s (15.6 mph). The BEM theory power prediction is more accurate for high wind speeds rather than low wind speeds. For low-range wind speeds ($U_\infty \leq 5$ m/s, 11.2 mph) the inefficiency of the generator is responsible for the high percentage of difference between the experiment and the theory while for high-range wind speeds ($U_\infty \geq 12$ m/s, 26.8 mph), stall delay phenomenon is mainly responsible for the inaccuracy. Moreover, neglecting radial flows in the theoretical analysis has a significant impact on the results for low and mid-range of wind speeds and leads to considerable errors in calculations.

Lanzafame et al. [14] compared the BEM results with the wind tunnel test results for a 10 m (33 ft) diameter rotor rated at 10 kW, with twisted blades and with a variable chord along the blade. They reported a maximum

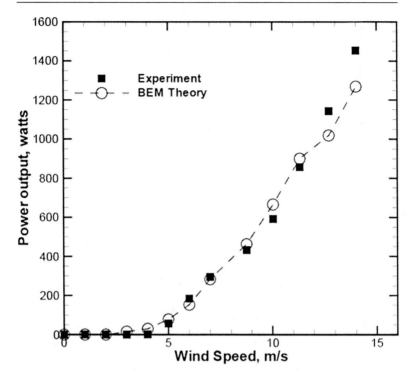

Figure 5.21. The theoretical and experimental power curve of the rotor for a wide range of wind speeds [17].

error of 9% for the range of 5 m/s $\leq U_\infty \leq$ 11 m/s (11.2 mph $\leq U_\infty \leq$ 24.6 mph) and a maximum error of 5.2% for the range of 11 m/s $\leq U_\infty \leq$ 14 m/s (24.6 mph $\leq U_\infty \leq$ 31.3 mph) in prediction of the power using the BEM theory including stall delay and tip-loss corrections. Therefore, the overall prediction of the BEM in the case of the smaller wind turbine (2.2 m, 7.2 ft diameter rotor) is not as accurate as the prediction of the theory for one particular turbine with a larger rotor (10 m, 32.8 ft in diameter). This discrepancy is mainly due to the effect of span-wise flows. As mentioned before, radial flows are important in the case of very small wind turbines and the difference between the power measured and the power calculated for this particular small wind turbine for low and mid-range wind speeds verifies the significant effects of neglecting radial flows in the analysis.

5.2 CONCLUDING REMARKS

The current state-of-the art in small-scale HAWT design, testing, and performance evaluation was described. The technology of small HAWTs is

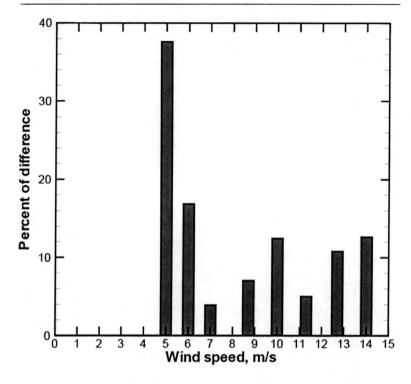

Figure 5.22. Percentage of difference between predicted power by BEM theory and the measured power [17].

under development. Until very recently, small HAWTs have been treated as a smaller version of grid-connected turbines. However, this vision is changing. Small HAWTs are usually deployed either in urban terrain or in the immediate vicinity of buildings. This has implications in terms of the wind environment to which these wind turbines are exposed. Instead of facing constant wind speed, they are exposed to complex, three-dimensional wind fields due to the built environment. This has a direct impact on both their performance and their resilience to wind turbulence. From the performance point of view, it is necessary to modify the theories in order to accommodate for the three-dimensionality effects associated with smaller-scale wind turbines. Furthermore, the blade design/selection must be based on increasing the efficiency of the machine while decreasing the noise. From the resilience point of view, new concepts need to evolve in the near future.

REFERENCES

[1]. Wilson, R.E., Lissaman, P.B.S., & Walker, S.N. (1976). Aerodynamic performance of wind turbines. Energy Research and Development Administration, ERDAINSF/04014-76/1.

[2]. Sahin, A.Z., Al-Garni, A.Z., & Al-Farayedhi, A. (2001). Analysis of a small horizontal axis wind turbine performance, *International Journal of Energy Research*, 25(6), 501–506.

[3]. Compositesworld Magazine (2012). Wind Power 2012 Report, http://www.compositesworld.com/.

[4]. Clausen, P.D., & Wood, D.H. (2000). Recent advances in small wind turbine technology, *Wind Engineering*, 24(3), 189–201.

[5]. Refan, M., & Hangan, H. (2012). Aerodynamic performance of a small horizontal axis wind turbine, *Journal of Solar Energy Engineering*, 134, 021013-1.

[6]. San Diego State University Foundation, Queen, Q. (2007). Build and test a 3 kw prototype of a coaxial, multi-rotor wind turbine, California Energy Commission, CEC-500-2007-111.

[7]. AeroVironment (2014). Architectural Wind, http:// http://www.avinc.com/engineering/architecturalwind1.

[8]. Selsam, D. (2003). "Latest Selsam Wind Turbine Sets World Record!" http://www.speakerfactory.net/wind_old.htm.

[9]. Yu, R. (2008). Airports go for green with eco-friendly efforts, http://usatoday30.usatoday.com/travel/flights/2008-09-16-green-airports_N.htm (accessed September 17, 2008).

[10]. http://www.compositesmanufacturingblog.com/2010/09/up-and-coming-turbines-series-part-ii/.

[11]. Manwell, J.F., McGowan, J.G., & Rogers, A. L. (2002). *Wind Energy Explained Theory, Design and Application*. Hoboken, NJ: John Wiley & Sons Ltd.

[12]. Hau, E. (2006). *Wind Turbines: Fundamentals, Technologies, Application, Economics*. London, UK: Springer.

[13]. Hansen, A.C., & Butterfield, C.P. (1993). Aerodynamics of horizontal-axis wind turbines, *Annual Reviews Fluid Mechanics*, 25, 115–149.

[14]. Lanzafame, R. and Messina, M. (2007). Fluid dynamics wind turbine design: critical analysis, optimization and application of BEM theory, *Renewable Energy*, 32(14), 2291–2305.

[15]. Betz, A. (1926). Windenergie und IhreAusnutzungdurchWindrniillen. Gottingen, Germany: Vandenhoeck and Ruprecht.

[16]. Burton, T., Sharpe, D., Jenkins, N., & Bossanyi, E. (2001). *Wind Energy Handbook*. Hoboken, NJ: John Wiley & Sons Ltd.

[17]. Refan, M. (2009). *Aerodynamic Performance of a Small Horizontal Axis Wind Turbine*, Master Thesis, The University of Western Ontario.

[18]. Himmelskamp, H. (1945). *Profile Investigations on a Rotating Airscrew*, PhD Thesis, Gottingen University, Germany.

[19]. Ronsten, G. (1992). Static pressure measurements in a rotating and a non-rotating 2.35 m wind turbine blade. Comparison with 2D calculations, *Journal of Wind Engineering and Industrial Aerodynamics*, 39, 105–118.

[20]. Snel, H. (1993). Sectional prediction of three-dimensional effects for stalled flow on rotating blades and comparison with measurements, Netherlands Energy Research Foundation ECN.

[21]. Glauert, H. (1935). *Airplane Propellers. Aerodynamic Theory*. Berlin: Springer Verlag (reprinted by Peter Smith, Gloucester, MA, 1976).

[22]. Wilson, R.E., & Walker, S.N. (1984). *Performance Analysis of Horizontal Axis Wind Turbines*. Corvallis, OR: Oregon State University.

[23]. Wood, D.H. (2004). Dual purpose design of small wind turbine blades, *Wind Engineering*, 28(5), 511–528.

[24]. Maalawi, K.Y., & Badr, M.A. (2003). A practical approach for selecting optimum wind rotors, *Renewable Energy*, 28, 803–822.

[25]. Selig, M.S. and McGranahan, B.D. (2004). Wind tunnel aerodynamic test of six airfoils for use on small wind turbines, Subcontractor Report, NREL/SR-500-34515.

[26]. UIUC. (2014). UIUC Airfoil data site, http://aerospace.illinois.edu/m-selig/ads.html.

[27]. Song, Q. (2012). *Design, Fabrication, and Testing of a New Small Wind Turbine Blade*, Master thesis, The University of Guelph.

[28]. Matsumiya, H., Ito, R., Kawakami, M., Matsushita, D., Lida, M., and Arakawa, C. (2010). Field operation and track tests of 1-kW small wind turbine under high wind conditions, *Journal of Solar Energy Engineering*, 132(1), 11002–11010.

[29]. Glauert, H. (1926). The analysis of experimental results in the windmill brake and vortex ring states of an airscrew, *ARCR R&M No.1026*.

[30]. Fitzgerald, R. E. (2007). *Wind Tunnel Blockage Corrections for Propellers*, Master Thesis, University of Maryland, College Park.

CHAPTER 6

NUMERICAL SIMULATIONS OF SMALL WIND TURBINES— HAWT STYLE

Jimmy C. K. Tong

Next, the author introduces the important topic of product certification and code compliance. He focuses on cases that are considered for numerical investigation of horizontal-axis wind turbine (HAWT) performance. Among the cases are different design environments with varying wind gusts, wind direction, and turbulence. Creation of the numerical mesh and application of the flow boundary conditions are discussed in detail. So too are the various turbulence models that modern computational simulators employ.

An emerging field in this area is the coupled interaction of the fluid and the solid object. This analysis approach, termed *Fluid Structural Interaction*, is finding use among research scientists as computer capabilities increase. Finally, the author discusses the types of information that are obtained from numerical simulation and how that information guides the design process.

6.1 INTRODUCTION

Given the rapid development of wind energy in small communities and residential buildings, the advancement of wind turbine design is crucial. The adoption of green energy is undoubtedly a worldwide trend, as more countries see the need to hand over a more sustainable earth to future generations. As the modern wind turbine has proven its effectiveness at least for the last 30 years, transforming this technology from a bigger scale to a smaller scale to overcome technical and cost challenges is commonly faced by the small wind turbine designers and manufacturers. Computational Fluid Dynamic (CFD) software is becoming a standard tool in the development of small wind turbines. This numerical simulation tool can assist designers and engineers to design and analyze turbines that are subjected to different wind conditions, allowing them to gain insight into turbine behaviors under various scenarios and to streamline design iterations to reach an optimal design with a shorter cycle. As a result, new wind turbine models can be developed for the intended market with higher performance and lower production cost in a shorter time.

This chapter is focused on the use of numerical simulation in the design of HAWTs. The two major categories of objectives for using numerical simulation are as follows:

1. Product certification and code compliance
2. Power generation and structural safety performance

In order to introduce a new wind turbine model into a market, most countries require or encourage product certification that is in compliance with local or international standards. To qualify for the certification, the wind turbine will need to pass the safety and standard performance testing and design verifications specified by the designated authority. This rigorous process is employed to ensure that the wind turbine will meet a certain quality standard and is safe to use, and the advertised performance is comparable to other products available in the market.

Design verification is part of the certification process, and numerical simulation can be helpful to demonstrate that the structural and safety requirements are met under the operating conditions. In general, numerical simulation uses the code-specified incoming wind condition to determine the wind loading acting on the blades and the turbine structure. Hence, the wind turbine can be designed to meet required material safety factors. This will be discussed in more detail later in this chapter.

Beyond design verification, the numerical simulation is also beneficial for designers to improve the performance of the wind turbine. In particular, a few computer codes are developed for blade design in order to capture kinetic energy from moving air. As the entire wind turbine comprises more components, the general CFD software is applicable for the complete system analyses. Details about the approach of numerical simulation, the analysis process with inputs and outputs, and the design iteration are discussed later in this chapter.

6.2 PRODUCT CERTIFICATION AND CODE COMPLIANCE

Product certification is a necessary process before a product can be introduced to customers. For most electronic products, the Underwriters

Laboratories (UL) and Conformité Européenne (CE) certifications are widely used around the world. Most countries adopt well-recognized international standards, while local codes still vary from place to place where the local authorities have jurisdiction over the local practice. For small wind turbine industries, the International Electrotechnical Commission (IEC) has published the following list of standards that covers the design and testing for wind turbine performance and safety:

- IEC 61400-2: 2006 Ed 2.0 Design requirements for small wind turbines [1].
- IEC 61400-11: 2012 Ed 3.0 Acoustic noise measurement techniques [2].
- IEC 61400-12-1: 2005 Ed 1.0 Power performance measurements of electricity-producing wind turbines [3].
- IEC 61400-14: 2005 Ed 1.0 Declaration of apparent sound level and tonality values [4].
- IEC 61400-21: 2008 Ed 2.0 Measurement and assessment of power quality characteristics of grid-connected wind turbines [5].
- IEC 61400-22 2010 Ed 1.0 Conformity testing and certification [6].
- IEC 61400-23: 2001 Ed 1.0 Full scale structural testing of rotor blades [7].

The standards of three countries are described next as examples:

In the United States, the American Wind Energy Association (AWEA) has the "Small Wind Turbine Performance and Safety Standard 9.1—2009," and a list of turbines that have received the Small Wind Certification Council approval can be found on its website [8].

Similarly, the British Wind Energy Association (BWEA) in the United Kingdom has its "Small Wind Turbine Performance and Safety Standard—2008," and those turbines that have received accreditation are listed on the Microgeneration Certificate Scheme (MCS) website [9].

In Denmark, the Danish Energy Authority (DEA) requires wind turbines to be certified under the Danish Certification Scheme, and those turbines that have received accreditation are listed on the Danish Certification website [10].

6.2.1 DESIGN LOAD CASES AND EXTERNAL CONDITIONS

While there are local code variations, most of the codes are very similar to the IEC 61400-2 [1]. Taking IEC 61400-2 as the basis for wind turbine design is commonly accepted by developers who intend their products for the global market. The standard specifies the design methodology, which includes simplified load equations, aeroelastic modeling, and mechanical load testing. Out of these methodologies, aeroelastic modeling is the best option for balancing accuracy and development time. Numerical simulation is a good tool to assist in this process, and compliance with the standard will drive how the numerical simulation should be constructed. The standard lists a set of design load cases (DLC; see Table 6.1), which in turn yields design loadings that generate the limit states. Then, the aeroelastic modeling of the wind turbine design is used to verify the safety operations under the specified cases.

In Table 6.1, seven basic design situations are covered and one or more DLC are specified for each situation. For each DLC, a wind condition and the type of analysis required (F for fatigue loading and U for ultimate loading) are prescribed. The several wind conditions used in these DLC are as follows and the details of each wind profile can be found in Chapter 6 of the standard IEC 61400-2 [1]:

ECD: Extreme coherent gust with direction change
ECG: Extreme coherent gust
EDC50: Extreme direction change with recurrence period of 50 year
EOG1: Extreme operating gust with recurrence period of 1 year
EOG50: Extreme operating gust with recurrence period of 50 year
EWM: Extreme wind speed model
NTM: Normal turbulence model
NWP: Normal wind profile model

Table 6.1. Set of Design Load Cases for aeroelastic models from IEC 61400-2.

Design situation	DLC	Wind condition		Other conditions	Type of analysis
1) Power production	1.1	NTM	$V_{in} < V_{hub} < V_{out}$ or $3V_{ave}$		F, U
	1.2	ECD	$V_{hub} < V_{design}$		U
	1.3	EOG$_{50}$	$V_{in} < V_{hub} < V_{out}$ or $3V_{ave}$		U
	1.4	EDC$_{50}$	$V_{in} < V_{hub} < V_{out}$ or $3V_{ave}$		U
	1.5	ECG	$V_{hub} = V_{design}$		U

(*Continued*)

Design situation	DLC	Wind condition		Other conditions	Type of analysis
2) Power production plus occurrence of fault	2.1	NWP	$V_{hub} = V_{design}$ or V_{out} or $2,5\ V_{ave}$	Control system fault	U
	2.2	NTM	$V_{in} < V_{hub} < V_{out}$ V_{el}	Control or protection system fault	F, U
	2.3	EOG_1	$V_{in} < V_{out}$ or $2,5\ V_{ave}$	Loss of electrical connection	U
3) Normal shut down	3.1	NTM	$V_{in} < V_{hub} < V_{out}$		F
	3.2	EOG_1	$V_{in} = V_{out}$ or $V_{max,\ shutdown}$		U
4) Emergency or manual shut down	4.1	NTM	To be stated by the manufacturer		U
5) Parked (standing still or idling)	5.1	EWM	$V_{hub} = V_{e50}$	Possible loss of electrical power network	U
	5.2	NTM	$V_{hub} < 0,7\ V_{ref}$		F
6) Parked and fault condition	6.1	EWM	$V_{hub} = V_{el}$		U
7) Transport, assembly, maintenance and repair	7.1	To be stated by the manufacturer			U

The author thanks the International Electrotechnical Commission (IEC) for permission to reproduce information from its International Standard ICE 61400-2 ed 2.0 (2006). All such extracts are copyright of IEC, Geneva, Switzerland. All rights reserved. Further information on the IEC is available from www.iec.ch. IEC has no responsibility for the placement and context in which the extracts and contents are reproduced by the author, nor is IEC in any way responsible for the content or accuracy therein.

The aforementioned eight different types of wind conditions describe specific wind scenarios. In different combinations with the DLC, the resulting loading for the wind turbine can be determined. These wind conditions describe, in general, the magnitude, turbulence, direction, wind shear, and time variation of these parameter changes. Furthermore, the behaviors of these wind conditions are different when the recurrence period is longer. For instance, a higher magnitude of extreme wind can occur when the recurrence period is 50 year as compared to extreme winds that recur yearly.

Together with the required DLC and the corresponding specified wind conditions, these form the inputs to the numerical simulations. The next section of this chapter further discusses how these inputs are incorporated into the analyses.

6.3 APPROACH TO NUMERICAL SIMULATION OF HORIZONTAL-AXIS WIND TURBINE

With the code and standard, which cover the certification process and define the safety and performance requirements of a wind turbine, numerical simulation is a common tool for the developers to properly assess the design and to effectively make design improvement through the iterative product development cycle. Out of the various important components of a wind turbine, like the blades, generator, brakes, controls of torque, pitch and yaw, and the tower, the numerical simulation discussed in this chapter only covers the mechanical aspect of the design. Other aspects, such as electrical and electromagnetic aspects, are not discussed.

As the blades and the rest of the turbine system (rotor, nacelle, and tower) are components that directly interact with the wind, numerical simulation for mechanical/structural loading and power generation (through transforming the kinetic energy from moving air into rotating mechanical energy for the generator to turn into electricity) can be determined through aeroelastic modeling and CFD modeling. For aeroelastic modeling, the Blade Element Momentum (BEM) theory has been commonly employed for blade design and airfoil selection. BEM allows simple calculations of steady loads, thrust, and power for different settings of wind speed, rotational speed, and pitch angle. Beyond steady calculations, and with the advancement of computation power, CFD is becoming the standard tool applied to wind turbines subjected to complex flow situations. Designers and engineers are still using both tools, independently or in combination,

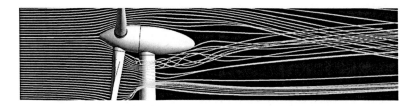

Figure 6.1. Streamline patterns from a small horizontal-axis wind turbine by CFD simulation.

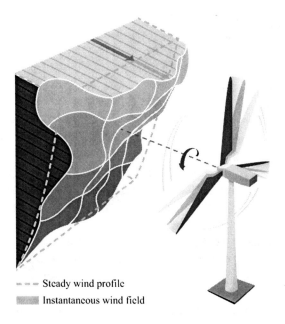

– – – Steady wind profile
▓ Instantaneous wind field

Figure 6.2. Schematic of wind profile.

for the turbine design. Figure 6.1 shows streamline patterns from a typical CFD simulation for a conventional three-bladed HAWT.

6.3.1 INPUTS TO NUMERICAL SIMULATIONS

As mentioned in the last section, the requirements specified by the standards provide most of the inputs to the numerical simulations. For instance, a visual of the wind profile acting on the wind turbine that describes a case for simulation is illustrated in Figure 6.2. The inputs can be categorized as follows:

External wind conditions: wind speed, direction, shear, turbulence, temperature, and density

Turbine geometrical specifications: blade design, nacelle dimension and weight, hub height, material properties, and tower design

Turbine operational specifications: revolutions per minute (RPM), pitch angle, and brake torque

6.3.2 PROPER SELECTION OF SOLUTION DOMAIN

In formulating the numerical simulation, it is recognized that not only are there geometrical complexities to be considered, but also relevant fluid flow processes that take place in the surrounding environment of the wind turbine. Furthermore, the flow is inherently three-dimensional. Considerable care is required to construct a mesh that would properly resolve these issues. Numerical simulation literature often states that once the conditions are specified on the boundaries of the chosen domain, the flow patterns are immediately affected by the features inside the solution domain. However, this phenomenon is a strong sign that the specified boundary conditions might not be appropriate. A recommendation for eliminating this type of error is to extend the solution domain so that the prescribed flow pattern remains for a while before being altered by the modeled features. Therefore, using an extended solution domain in order to obtain results of high accuracy is deemed necessary.

The observations set forth in the preceding paragraphs merely illustrate the importance of the proper selection of the boundaries of the solution domain. Moreover, sufficient upstream and downstream spaces are needed to allow proper specification of the boundary conditions. The solution domain can be considered large enough when the boundary conditions specified are able to maintain the original characteristics for a while before being affected by the presence of the wind turbine. In other words, the arbitrarily chosen virtual boundaries should not have any impact on the flow results of interest. This idea follows the same principle of designing a wind tunnel in such a way that the edge effects are minimized.

6.3.3 MESHING AND NODE DEPLOYMENT OF SOLUTION DOMAIN

The key to attaining numerical accuracy in simulation studies is to construct an appropriate mesh. Most commercial CFD software is based on the finite-volume method of discretizing the solution domain. The accepted approach for seeking mesh-independent solutions is to systematically refine the mesh. This approach, although widely accepted, possibly has invalid indications. For example, if the elements that compose the mesh are all diminished by the same multiplicative factor without accounting for regions of high gradients, mesh independence may appear to exist whereas, in fact, proper attention to high-gradient zones may indicate a different conclusion. Another approach, although less systematic, is to concentrate on regions of high gradients. Here, a combination of the two approaches is recommended to assess mesh independence.

In view of the previous paragraph, studies of mesh independence can be performed by comparing the simulation results of coarser and finer meshes based on the aforementioned approach. Along the same lines, the results of interest and of high-gradient zones can be set as figures of merit for the attainment of mesh-independent solutions.

Attention is now turned to two types of CFD analysis—(i) blade and rotor level and (ii) entire turbine system level. The first analysis may have an advantage of focusing on the blade performance, whereas the second one can investigate the system response for the entire turbine. For the first level of CFD analysis, which is conducted on the blade only or on the blades with a rotor, a sample mesh with the blade and rotor is shown in Figure 6.3. The solution domain contains the surfaces of a single blade and a portion of the rotor. As this is a three-bladed HAWT application, the fluid domain is chosen to be one-third of the entire circle. With sufficient upstream and downstream fluid domain included, the entire mesh forms like a pie as shown in Figure 6.3. Mesh refinement is performed near the

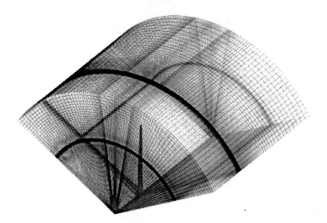

Figure 6.3. Meshing for wind turbine blade and rotor CFD simulation.

Source: Reproduced from Bazilevs et al. [11].

blade and rotor surfaces to resolve the boundary layer of the fluid on the leading and trailing edges and tip of the blade. With only one-third of the domain modeled, it is possible to obtain a solution using a numerical technique of periodicity. The periodic boundary conditions can be used when the flows going out through one boundary reappear as the flows going in through the opposite boundary. If the flow pattern is too complex for the periodicity to be appropriate, the same mesh can be repeated two more times every 60°. When the complete face of the blade and rotor is modeled, then the usual CFD analysis without using periodicity can be conducted.

The other type of analysis covers the entire wind turbine system, which includes all the blades, rotor, nacelle, and tower structure. This analysis can be useful to simulate a real site condition or a wind tunnel test situation. A sample mesh for this simulation is shown in Figure 6.4.

As seen in Figure 6.4, the solution domain is chosen for a wind tunnel. In both the wind tunnel and real site location, the recommended distances to form the solution domain are 2–5 times the rotor diameter as the upstream distance, 10–20 times as the downstream distance, and 1–3 times from the edge as the distance from both sides. Mesh refinement is done near all surfaces to allow finer resolution of the flow pattern to be developed within the boundary layer. In particular, special attention on mesh refinement is required near the leading and trailing edges and the tip of the blades.

Since wind turbines are a type of rotating machinery, a numerical simulation technique using a rotating frame of reference is appropriate

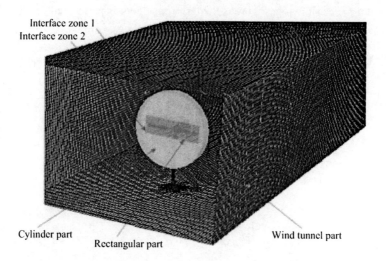

Figure 6.4. Meshing for entire wind turbine system CFD simulation.
Source: Reprinted from Moa et al. [12].

to simulate the moving blades and rotor. The fluid inside this rotating domain is considered to be moving with the frame, and this simplifies the computation and allows faster completion of the calculation. As shown in Figure 6.4, the surfaces of blades and rotor are meshed in the rectangle part and the rectangle part is inside the cylinder part, which causes the full rotation of the blades and rotor. Both the rectangle and cylinder parts are considered moving domains, whereas the wind tunnel part is treated as a stationary domain. The sliding mesh technique is employed and the fluid interfaces are defined between the aforementioned parts. Hence, interface zone 1 is located between the rectangle and cylinder parts, and interface zone 2 is located between the cylinder and wind tunnel parts. Although this technique may be computationally demanding, it is the known method for achieving high accuracy unsteady solutions for moving and stationary domains that comprise multiple frames of reference.

6.3.4 BOUNDARY CONDITIONS

The boundary conditions are specified by the standard, and the inputs described in the previous section are to be translated as the boundary conditions. Beyond the conditions from product certification and code compliance, conditions from the actual site location can also be used in the simulation.

6.3.5 NUMERICAL SOLVER

Focus is now turned back to CFD simulation to further determine the next step for completing the analysis. The wind blows in open air; therefore, most flow situations are turbulent flows. Due to the complexity of the different flow patterns, numerous models are developed to balance the accuracy and speed of the computation. In addition to flow modeling, simulation with multi-physics is becoming more common for achieving better designs. One combined model worth mentioning is the fluid–structure interaction (FSI) model, which is discussed later.

6.3.6 TURBULENCE FLOW MODELS

As the development of computational power continues to increase and more complex flow patterns are able to be formulated, more turbulent flow models can be created, ranging from a few specialized generic types of flow patterns to more generalized codes that cover wide applications. The guiding principles of formulating the CFD code include the law of conservation of mass and Newton's second law of motion. In the specialized literature for fluid mechanics, Newton's second law for flowing fluid is sometimes called momentum conservation or, alternatively, the Navier–Stokes equations. Conservation of mass states that mass cannot be created or destroyed, while Newton's second law balances forces with changes of momentum. The following list of turbulent flow models is available through most commonly known software:

1. Reynolds-Averaged Navier-Stokes Simulation (RANS)
   Standard k–ε model
   RNG k–ε model
   Realizable k–ε model
   Standard k-ω model
   Shear–stress transport (SST) k-ω model
   Reynolds stress model (RSM)
2. Detached Eddy Simulation (DES)
   Spalart–Allmaras model
3. Large Eddy Simulation (LES)
   Smagorinsky model
   Algebraic Dynamic model
   Localized Dynamic model
4. Direct Numerical Simulation (DNS)

The aforementioned list covers four major groups of models, presented in the order of their complexity and their demand for computational resources. The first group of models is known as RANS, where the numerical formulation of the term "turbulent viscosity" is introduced in the Navier–Stokes equations. It is adjusted to suit different flow patterns and account for the effects of turbulence. As each of the specific models in this group uses different methodologies to solve equations in different ensemble-averaged forms, different equations of turbulent viscosity are used in each of these models.

The second group is DES, which uses a hybrid method combining the RANS and LES approaches to treat near-wall regions and the bulk flow respectively.

The third group is LES, where the momentum and energy transfer of large energy-carrying turbulent structures are computed exactly by using the governing equations, while the effect of the sub-grid scales (SGS) of turbulence is modeled or approximated. LES applies low-pass filters according to turbulence theory and available computational resources to separate the "small" length and time scales in the velocity field. It is known to be more accurate than RANS but less demanding than DNS.

Lastly, the DNS approach resolves all scales of turbulence by solving the Navier–Stokes equations directly without any turbulence modeling. DNS requires a very fine mesh; hence, it has a very high demand for computational resources in order to achieve high accuracy.

6.3.7 FLUID–STRUCTURE INTERACTION

In the context of wind turbine applications, FSI is the interaction of the deformable structure, like the blade and tower, with the surrounding fluid flow. For instance, the blade deforms from wind loading and the deformed shape changes the flow, which in turn affects the performance. This interaction mostly affects the blade structure design and aerodynamic performance. To carry out the analysis of FSI, it is recommended to perform a two-way iterative loop with a mapping of calculated results between a CFD model and another Finite Element (FE) model. For instance, the pressure results on the blade are first obtained in the CFD model, and then mapped on the FE model with the original shape. After the FE calculation, the deformed shape is passed back to the CFD model to form a new geometry for flow calculation. The process continues until convergence is found.

6.3.8 SOFTWARE

Numerical simulation using CFD is a widely adopted tool in both research and industry. Many software packages are available through commercial licenses and open sources. Many of these codes are developed for generic applications and some of these are tailor-made for wind turbine design. To name a few brands as reference, ANSYS/Fluent/CFX, Star-CD, Pointwise, ACUSIM, SimPack, and FAST are among the most common ones. Besides CFD, a specially designed code, Bladed, which uses BEM, is also a reliable tool and is widely employed by wind turbine developers in the industry.

6.4 POWER GENERATION AND STRUCTURAL SAFETY PERFORMANCE

To achieve higher performance of the wind turbine design, numerical simulation is an effective tool to provide an accurate and quick assessment on the design under the specified environmental conditions. Using the inputs as prescribed in the codes and standards and determined by operational design parameters, the turbine design is modeled in the properly chosen solution domain using the inputs as boundary conditions. With a fine enough meshing determined by a mesh-independence study to ensure the simulation quality and the appropriate flow solver, the results can be extracted.

Among the vast amount of data in a complete set of simulation runs, the key results that drive the turbine design are pressure and thrust coefficients of the blade, the power curve and power coefficient of the wind turbine, the loading experienced at the blade, the drive train, the tower of the turbine, and the wake losses created by the wind turbine. Other resulting factors associated with each design iteration are the material used and cost.

From the numerical simulation that focuses on the details of the blade design (i.e., the selection of airfoil, blade length, chord length, twist angle, relative thickness, and pitch axis), the calculated results of the pressure and thrust coefficients are helpful to designers. The coefficients can be used to give designers the insights to balance the power that can be captured by the blade and the structural loading acting on the blade. Based on trade-off analysis, an optimal design is reached for the specified condition. Figure 6.5 shows an example of calculated pressure coefficients along a blade from numerical simulation and compares the values from a wind tunnel test.

For the overall wind turbine performance, the same approach is used to evaluate the "effectiveness" of the design; it balances between the power generation and the structural integrity for safe operation. An example of a typical power curve is shown in Figure 6.6. In the figure, the resulting power curves for models #1–#3 illustrate different design approaches, some with higher power generation while others more focused on reducing loading on the turbine structure. As for the structural aspect, the code and standard have suggested a preset format for the present load data of the blade, drive train, and the tower of the turbine. The format helps to organize a large amount of data and allows attention to be focused on the limiting cases for design verification and improvement. An example of the format is shown in Table 6.2.

Finally, by analyzing the performance of each design iteration, improvements are made possible by changing the design parameters. To speed up and shorten the design iteration cycle, many optimization algorithms are available to run in conjunction with the numerical simulations. Many successful products and projects are examples of this process.

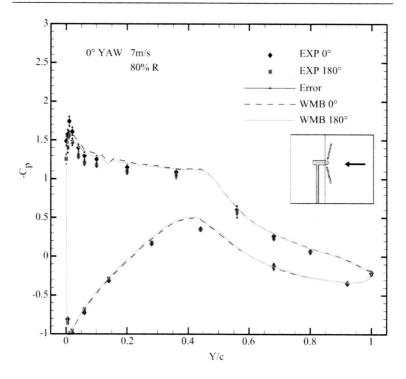

Figure 6.5. Pressure coefficient (C_p) on the blade surface along the entire blade length.

Source: Reproduced from Gómez-Iradi, Barakos, and Munduate [13].

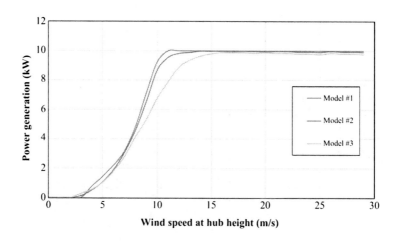

Figure 6.6. Power curves for a typical 10 kW horizontal-axis wind turbine.

Table 6.2. Suggested format to present load results for extreme (top) and fatigue (bottom) loads.

Results of the extreme load evaluation at a component									
Design Load case	Safety factor	Fx [kN]	Fy [kN]	Fz [kN]	Fres [kN]	Mx [kNm]	My [kNm]	Mz [kNm]	Mres [kNm]
DLC with Fx Min									
DLC with Fx Max									
DLC with Fy Min									
DLC with Fy Max									
DLC with Fz Min									
DLC with Fz Max									
DLC with Fres Min									
DLC with Fres Max									
DLC with Mx Min									
DLC with Mx Max									
DLC with My Min									
DLC with My Max									
DLC with Mz Min									
DLC with Mz Max									
DLC with Mres Min									
DLC with Mres Max									

Results of the fatigue load evaluation at a component

Inverse S-N slopes	Fx [kN]	Fy [kN]	Fz [kN]	Mx [kNm]	My [kNm]	Mz [kNm]

6.5 CONCLUDING REMARKS

To deliver a new HAWT to the market for small-power applications with a shortened design cycle, numerical simulation is a necessary tool for designers and engineers. Numerical simulation allows developers to gain the insights of the turbine design and to evaluate the effects of varying the design parameters in a much shorter time than using traditional techniques. By running the code-required load cases under the specified wind conditions, the simulated results can be used to provide evidence that a wind turbine is in compliance with the code and standard; therefore, the first objective is achieved. Going beyond the code and standard, the other objective of employing numerical simulation is to achieve an optimized design. Designing wind turbines is a complex process and many parameters affect both power generation and structural performance. Numerical simulation allows turbine developers to balance multidimensional targets with multifaceted constraints.

While there are attractive and obvious reasons to use numerical simulation to assist the design process, special care is required to formulate the approach to achieve accurate and meaningful results. Many steps have been explained in this chapter, starting with the proper selection of the solution domain, the mesh-independence study for mesh quality, the correct translation of the design inputs as boundary conditions, and ending with the choice of an appropriate numerical solver to obtain the results. Thoughtful evaluation of the selected outputs is necessary to identify the degree of impact by each of the design parameters. Through the design iteration cycle, the determination of a possible range of the design parameters affects the final attainment of the optimal design and the time required for this process. Therefore, careful planning for each of the aforementioned steps is recommended.

6.6 ACKNOWLEDGMENTS

The author would like to express appreciation to Janet Pau and Wai Heng Ken Cheng for their support for this chapter. In addition, the author would like to thank professor Ephraim Sparrow for his friendship and inspiration for professional excellence.

REFERENCES

[1]. International Electrotechnical Commission; IEC 61400-2: 2006 Ed 2.0 Design requirements for small wind turbines.
[2]. International Electrotechnical Commission; IEC 61400-11: 2012 Ed 3.0 Acoustic noise measurement techniques.
[3]. International Electrotechnical Commission; IEC 61400-12-1: 2005 Ed 1.0 Power performance measurements of electricity producing wind turbines.
[4]. International Electrotechnical Commission;IEC 61400-14: 2005 Ed 1.0 Declaration of apparent sound level and tonality values.
[5]. International Electrotechnical Commission; IEC 61400-21: 2008 Ed 2.0 Measurement and assessment of power quality characteristics of grid connected wind turbines.
[6]. International Electrotechnical Commission; IEC 61400-22: 2010 Ed 1.0 Conformity testing and certification.
[7]. International Electrotechnical Commission;IEC 61400-23: 2001 Ed 1.0 Full scale structural testing of rotor blades.
[8]. http://www.smallwindcertification.org.
[9]. http://www.microgenerationcertification.org/mcs-consumer/product-search.php.
[10]. http://www.dawt.dk/DK/Godkendte_small_WT.htm.
[11]. Bazilevs, Y., Hsu, M.C., Akkerman, I., Wright, S., Takizawa, K., Henicke. B., Spielman, T., and Tezduyar, T.E. (2011). 3D simulation of wind turbine rotors at full scale. Part I: Geometry modeling and aerodynamics. *International Journal for Numerical Methods in Fluids*, 65, 207–235.
[12]. Moa, J.O., Choudhrya, A., Arjomandia, M., and Lee, Y.H. (2013). Large eddy simulation of the wind turbine wake characteristics in the numerical wind tunnel model. *Journal of Wind Engineering and Industrial Aerodynamics*, 112, 11–24.
[13]. Gómez-Iradi, S., Barakos, G.N., and Munduate, X. (2010). A CFD investigation of the near-blade 3D flow for a complete wind turbine configuration. 2010 European Wind Energy Conference & Exhibition, EWEC, Warsaw.

CHAPTER 7

CASE STUDIES OF SMALL WIND APPLICATIONS

M.H. Alzoubi

In the final chapter, case studies are presented where the application of small-scale wind power solutions is evaluated. In total, eight cases are described that range from small to medium size. They include both on- and off-grid applications in developed and developing locations around the globe. The author provides a background for each case study that motivates the wind-power application. Design specifications are given and a match between the design and the energy requirement is made. Economic analysis is included when available. Readers can use the experience outline in this chapter to assess the potential of future small-scale wind installations.

7.1 INTRODUCTION

Small wind turbines are defined, in this chapter, as turbines with a capacity rating of less than 100 kW. The turbine size is determined according to the application and the required loads it will feed. Therefore, off-grid applications usually employ small wind turbines, whereas on-grid applications rely on larger turbines. A wide range of small turbine models are now commercially available for different applications including homes, schools, medical centers, banks, clubs, farms, and industrial facilities. As a result, the small-turbine market is now experiencing rapid growth supported by an unstable state of oil prices worldwide.

The case studies considered here will include off-grid and on-grid wind energy systems, using different turbine sizes feeding various loads. For each considered case, the rationale for wind energy selection as a power source is discussed. The case descriptions include the wind speed, the turbine size, the load type, the annual turbine output, the public interest in the project, the power savings, and the payback period. Finally, the cultural and social dimensions of installing wind turbines are highlighted for some of the cases studied.

7.2 OFF-GRID SMALL WIND SYSTEMS

Off-grid applications, which solely rely on diesel generators as a main energy source, offer a huge opportunity for the deployment of wind turbines and have a great potential for reducing carbon emissions. For remote locations, the cost of being connected to a grid is often not economically viable; so users in those locations incur high fuel and transportation costs. Therefore, the introduction of small wind turbines for off-grid systems as the primary energy source, together with batteries as a storage medium, can have a remarkable effect on reducing costs and pollution.

7.2.1 CASE STUDY 1: OFF-GRID WIND SYSTEM WITH 2 KW TURBINE

7.2.1.1 Background

In June 2012, a 2 kW wind turbine was installed at Rasoon tourism camp in Ajloun province, 70 km to the north of Amman, the capital of Jordan. The camp was built at the top of a windy hill (mean wind speed: 6.8 m/s, 15 mph), with the help of U.S. Aid and the European community. The camp was far from the grid and the owner was struggling with an old gasoline generator to power his 20-tent camp with electricity. He also needed to provide a reliable electric supply for the personal equipment of the visitors, who typically stay for several days in the camp. The main loads were lighting, computers, small electric device rechargers, and a refrigerator. Figure 7.1 shows a view of the camp area.

The camp owner contacted the author for consultation concerning the selection and installation of a wind turbine in the camp. The project was started by conducting a survey of the loads of the camp and the mean wind speed at the site to determine the required size and optimal location for the wind turbine. Then, detailed specifications for the required turbine, charger, inverter, controller, dump load, and batteries were set.

7.2.1.2 Turbine Specifications

The installed turbine has the specifications as listed in Table 7.1.

Figure 7.1. View of the Rasoon camp.

Table 7.1. System components and specifications

Item	Specifications	Item	Specifications
Turbine type	ZH2 kW	Maximum wind speed	50 m/s (112 mph)
Rotor diameter	3.6 m (12 ft)	Working voltage	DC24/48 V
Material/no. of blades	Reinforced fiber glass, 3	Generator type	3 phase, PM
Rated/maximum power	2 kW/2.5 kW	Charging	Constant voltage
Rated wind speed	10 m/s (22 mph)	Speed regulation	Autofurl
Start-up wind speed	2.5 m/s (5.6 mph)	Tower height	9 m (30 ft)
Working wind speed	3–25 m/s (6.7–56 mph)	Life time	15 y

The tower and anchoring system were manufactured in the local market to reduce the transportation cost. The batteries and connection wiring were also purchased from local vendors to reduce the total cost of the system. During the installation, several volunteers actively participated in the turbine assembly, preparing the foundation, and erecting and fixing the tower. The total cost of the system was $5,200. Figure 7.2 shows a view of the turbine and Figure 7.3 illustrates the connection diagram of the wind system.

7.2.1.3 Annual Energy Output

The annual energy output of the turbine is expected to be more than 4,300 kWh. This will exempt the camp owner from paying $3,000 for gasoline.

Figure 7.2. View of the 2 kW wind turbine.

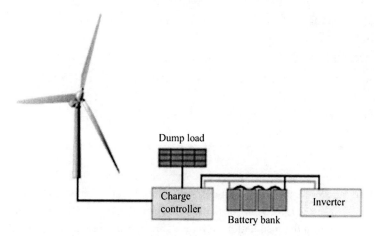

Figure 7.3. The connection diagram of the 2 kW wind system.

The payback period for the system is less than 2 y. With this small turbine, the owner can take greater advantage of the remote location of the camp and harness the power of the wind.

7.2.1.4 Cultural Aspects

It is worth mentioning here that most of the camp visitors were interested in this project and for many of them this was the first time they had seen a wind turbine. The visitors and camp neighbors started to ask about its advantages and considered installing similar ones for their own use. This will, inevitably, enhance the environmental values and improve the natural respect for this beautiful area.

7.2.2 OFF-GRID WIND SYSTEM FOR WATER PUMPING

7.2.2.1 Background

The use of wind energy for water pumping is an old application. Traditionally, water has been pumped to drain marshes, produce salt from seawater, and irrigate agricultural land. As windmill technology improved, water pumping applications grew in number and diversity. The introduction of the American wind pump in the late nineteenth century marked the arrival of mass availability of wind-pumping technology. These systems were able to operate for long periods of time, even if left unattended. They were used to supply water for domestic use, livestock, and steam engine operations throughout the United States. Millions of American wind pumps have been produced and installed around the world. It is thought that over a million are still in use today. However, this traditional wind pumping technology did not come without its drawbacks. Due to their simple multiblade design, poor material selection, and inherent mechanical flaws, American wind pump output was only 4%–8% of the energy available in the wind. So, despite the advantages of this type of wind turbines, they have several drawbacks as reported in the literature [1,2].

As an alternative to the American wind pump, electrical wind pump systems can be designed with lower solidity rotors than traditional wind pumps. They are also capable of generating higher tip speed ratios and producing more power. Although they require a higher starting wind speed, wind turbines are twice as efficient at extracting wind energy compared to traditional wind pumps. They have fewer moving parts that reduces the need for maintenance, and their costs are competitive with those of traditional wind pumps [2]. Since electricity can be transmitted long distances through a simple power line, water-pumping wind turbines can be placed far away from the site of water pumping, at locations where the wind resource is greater. They are capable of responding to a wide range of wind speeds without a significant reduction in efficiency and can do this affordably by using variable speed generators that do not require complex power rectification systems.

7.2.2.2 Electrical Wind Pump Design

The main components of an electrical wind pumping system include the turbine rotor, tower, electrical generator, motor, pump, as well as electrical wiring. Variable speed systems are ideal for such applications since they require no gearbox and can operate at high rotor speeds with less structural loading on the turbine. They are considered more reliable and less costly to maintain. This is important in off-grid applications where maintenance can be difficult [1].

It is worth mentioning here that an appropriate pump must be selected to match the expected output conditions of the motor. In many cases, pumps are designed with built in multistage motors designed to operate over a range of input conditions. In addition to the selection of an appropriate wind turbine and pumping system, many other aspects need to be taken into consideration when designing a wind pumping system. These include the construction of a well or storage reservoir from which water is to be pumped, a storage tank at the desired water output location, and all necessary plumbing.

A location must be selected for the turbine that will give the best wind resource. An appropriate tower height must be selected and wiring must connect the turbine to the pump motor [3]. A schematic diagram, showing common electrical wind pumping systems, is presented in Figure 7.4.

Several case studies of off-grid wind pumping are available in the literature. Applications range from water distribution to irrigation and watering livestock. In this section, three different cases are reported for various uses of wind pumping in developing countries [3].

7.2.3 CASE STUDY 2: NAIMA WATER NETWORK, MOROCCO

A 10 kW system with a rotor diameter of 7 m (23 ft) has been installed to drive an electric water pump in Ain Tobin, Morocco. This place is located in a semiarid area where water supplies are often scarce. Children are

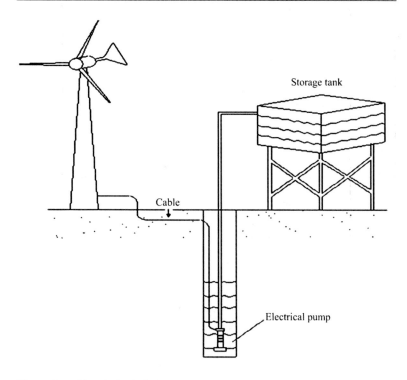

Figure 7.4. Electrical wind pump.

typically given the task of collecting and handling water from a source located more than 4.5 km (2.8 miles) from the community housing. The houses in this location and at the water source were not connected to the electrical grid. Therefore, the decision to employ this wind pumping system as a replacement of an old diesel installation was very significant. The operation and maintenance costs of the old diesel generator were intolerable for that community, particularly because of the high cost of diesel fuel.

The annual mean wind speed in the area was 4.11 m/s (9.19 mph) and the lowest wind speed was recorded in October to be 3.6 m/s (8.0 mph). It is expected that the wind energy system can produce 105% of the target delivery for AinTolba and Elhamra during the lowest wind speed month of October.

In this system, the turbine rotor drives an alternator that produces a three-phase alternating current that varies in voltage and frequency with wind speed. The electrical output is used to drive three-phase induction motors on the pumps at variable speed. The wind system was designed to fill a storage tank, where the people can collect water. When wind speeds

154 • SMALL-SCALE WIND POWER

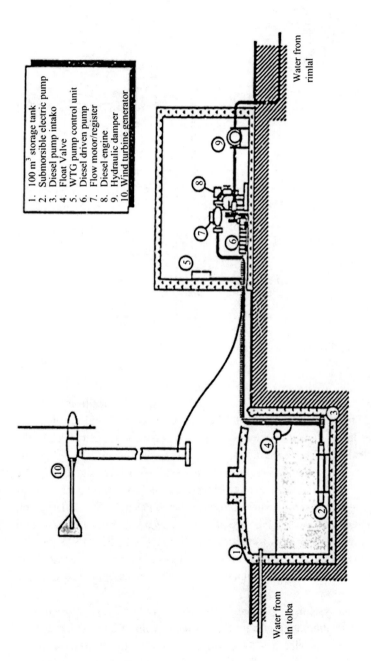

Figure 7.5. Naima project equipment layout.

1. 100 m³ storage tank
2. Submersible electric pump
3. Diesel pump intake
4. Float Valve
5. WTG pump control unit
6. Diesel driven pump
7. Flow motor/register
8. Diesel engine
9. Hydraulic damper
10. Wind turbine generator

are sufficient, water is pumped from a spring-fed storage tank to a larger tank half a kilometer away. The latter provides a service for four villages and 3,500 people. The system was installed in 1989 and has supplied water in excess to the community ever since.

The equipment layout is shown in Figure 7.5 [4].

7.2.4 CASE STUDY 3: B. LLASARIA, ALTIPLANO, BOLIVIA

Llasaria, in the southeast of Bolivia, is a community of about 1,000 people. The inhabitants obtain drinking water from a stream contaminated with cholera-causing bacteria. The climate in the area is cool and humid to semiarid and even arid, with mean annual temperatures that vary from 3°C (37°F) near the western mountain range to 12°C (54°F). Most of the year, the weather is cool, windy, and dry.

In 1995, a 10 kW wind turbine was installed to pump water from a well, 1 km (0.62 miles) away. To maximize wind availability, the turbine was located on a hill 700 m (2,300 ft) from the well. It was capable of providing the community with about 35 m^3 (9,250 gal/day) of safe drinking water per day. This amount was enough for the people to fulfill their needs from this water [4]. The average annual energy production at 5 m/s (11 mph) wind speed is 13,200 kWh.

There was no access to a grid when the turbine was installed and, therefore, it was operated as an off-grid system. However, now the community is serviced with electrical energy and the system can be easily converted into an on-grid system if the local regulations allow for that.

7.2.5 CASE STUDY 3: C. HEELAT AR RAKAH, OMAN

In December 1996, the first wind-powered, electric water-pumping system in the country of Oman was successfully installed at the Ministry of Water Resources camp in Heelat Ar Rakah, a remote location some 900 km (560 miles) south of the capital, Muscat. A weather-monitoring station comprising temperature, wind speed/direction, humidity, and barometric pressure equipment was also installed to investigate the relationship between weather conditions and power output. Both systems are still in operation and are providing valuable data for analysis.

A 10 kW wind turbine was selected. It was installed on a 24 m (79 ft) tower and equipped with a submersible water pump. The relationship among wind speed, power output, and pumped water was continuously measured. These values were compared with the designed output values provided by the system manufacturer. The mean annual wind speed at 10 m (33 ft) height in the area is 5.7 m/s (13 mph). The turbine provides 70 m (230 ft) vertical lift and has roughly 30 m (98 ft) of losses, giving a total head of 100 m (330 ft). A pumping rate of 30 m^3/day (8,000 gal/day) was intended to meet irrigation demands at the camp. After installation, it was found that the system operates as expected and meets the water requirements more than 80% of the time during the months of interest. The rest is provided by a backup diesel generator system [5]. Figure 7.6 shows a picture of this turbine.

7.2.6 CASE STUDY 4: OFF-GRID WIND SYSTEM FOR RESIDENTS IN ALASKA, USA

Appalachia Wales, Alaska, a small community of about 160 persons, primarily Inupiat, is situated at the end of the Seward Peninsula on the Bering Strait. On a clear day, it is easy to see all the way to Russia from this northern outpost of the United States as reported in Ref. [6]. Wales is an ideal site to examine the challenges and benefits of operating small wind turbines in a remote location and harsh weather.

In 2002, a wind–diesel hybrid research and development system was installed. The project was jointly funded by the federal Environment Protection Agency, the United States Department of Energy, the State of

Figure 7.6. Wind turbine installed at Heelat Ar Rakah.

Figure 7.7. The 65 kW wind turbine installed in Appalachia, Wales, Alaska.

Alaska Energy Authority, and the Alaska Science and Technology Foundation. This innovative hybrid project incorporated two 65 kW wind turbines, 3 diesel generators, 2 electric dump loads, a rotary power converter, and a battery bank. The system is owned by Kotzebue Electric Association (KEA), a rural Alaskan public utility. Figure 7.7 illustrates a view of the employed wind turbine [6].

Wales has among the best wind resources in the world—Class 7—with an average wind speed of nearly 8.9 m/s (20 mph). Extreme winds here on this peninsula are rare; so it was expected to be a very productive site. Several times each winter, however, there are storms out of the south, where the wind approaches from over open water. These winds can cause the wind turbine blades to ice up, which can lead to turbine unavailability for days or even several weeks, depending on the ensuing weather [6]. Among the lessons learned from this project is that any future installation in a similar icing environment should include blades painted black to improve radiative heating and with a special antistick coating applied.

7.2.6.1 Public Acceptance

Local residents were generally not opposed to the project. Much of the populace in the Alaskan town was largely indifferent, interested only to the extent that the wind turbines would lower their electric bills. Some local residents have benefited from employment chances by the project during the construction and installation phases, but the ongoing requirement for local help was quite modest, as reported in the literature [6].

One particular advantage of this system is that any excess wind power above what is required to meet the primary village electric demand is sent to one or two electric boilers that were installed as part of the project. As a result of this arrangement, the wind turbines in this configuration not only reduce the amount of fuel used to generate electricity in the village, they also reduce the amount of fuel used for heating at a local school.

7.2.6.2 System Performance

Overall, this system has performed quite well. However, the system has suffered from periods of poor availability of the wind turbines, which has caused the fuel savings and the amount of diesel-off time to be much less than it could have been based on the superior local wind resource. In most cases, the problems have been relatively minor and were addressed with fairly small repairs or design modifications. The challenge of this situation is that Wales is very remote and is difficult and expensive to travel to. Being a technologically complex system, practically all problems

Table 7.2. Specifications of the off-grid system feeding the examined family houses

Item	Characteristics	Item	Characteristics
First wind turbine:		*Second wind turbine*:	
Rating, kW	4.5	Rating, kW	0.9
Rotor diameter, m (ft)	4.5 (15)	Rotor diameter, m (ft)	2.1 (7)
Battery charging	Typical	Battery charging	Typical
PV panel:		*Battery bank*:	
Rating, kW	1.44	Voltage rating, V	48 V DC
Inverter:		*Diesel generator*	
AC voltage, V	240	Rating, kW	12
No. of phases	1 phase		
Total power, kW	8		

have required a visit by a technician, sometimes from as far away as the National Renewable Energy Laboratory in Boulder, Colorado.

The host utility, the Alaska Village Electric Cooperative (AVEC), the owner/operator of the Wales power system, has saved a substantial amount of fuel since the system began operating while paying a considerable amount for plant upgrades. This project has highlighted, as have many others, the unique benefits and ingenuity needed to sustain renewable energy power systems including wind turbines in remote areas [6].

7.2.7 Case study 5: Off-grid Wind System for a Family House in Colorado, USA

This case study was an off-grid wind energy system designed and installed to provide power to a family of three persons living on 120 acres (half kilometer square) in rural Douglas County, Colorado, United States. The home is located near the Palmer Divide, an elevated region between Denver and Colorado Springs. The local topography of the area provides unusually energetic winds for eastern Colorado. The off-grid home of this family is powered by a hybrid wind and photovoltaic system with diesel backup [7].

7.2.7.1 System Configuration

The considered example house is fed from several renewable energy sources as described in Table 7.2.

The home has electric space heaters and hot water. Excess wind power has been adequate to serve these thermal loads. Only recently, a propane tank was added for backup. The major appliances in the home are a refrigerator, a washer, and an energy-saving drier. Due to the success achieved in this system, the house owner planned to add two more wind turbines in the future, 900 W and 1,000 W, to expand the system [7].

The 4.5 kW wind turbine is installed on the top of a hill about 90 m (300 ft) above the house. This location has higher wind speeds than those at the house, but also turbulence created as the wind flows up the slope of the hill. The tower height, formerly 13 m (43 ft), has been increased to 19 m (63 ft) in an attempt to position the wind turbine in smoother air.

The second turbine of 0.9 kW capacity is located on the hill slope within 30 m (100 ft) of the house as shown in Figure 7.8. The wind resource here is still quite good and turbulence is reduced. This turbine is less than 1 y old and has performed flawlessly to date. It is audible inside

Figure 7.8. The 900 W wind turbine on hillside above Douglas County home.

the home when it is furling in high winds above 13.4 m/s (30 mph). Its sound is like the propeller of a small airplane, whereas, at lower wind speeds, it is very quiet.

Finally, it was found that the connection to the local electric grid would cost the owner of this house about $12,000 [8].

7.3 ON-GRID WIND SYSTEMS

On-grid applications, which mainly rely on the electrical networks as the main energy source, have a great opportunity for deployment of wind turbines to reduce their electricity consumption from fossil fuels and to lower carbon emissions. In non-oil countries, the cost of electricity is frequently higher than that in regions with domestic petroleum supplies. Moreover, the rate of rise in the cost of electricity is proportional to the amount of consumption. Therefore, the introduction of wind turbines for on-grid applications to reduce fossil fuel expenditures can have a significant effect on reducing costs and pollution.

In areas that experience good wind, customers are encouraged to install on-grid wind systems to benefit from this freely available energy. Usually the payback periods for such projects are relatively short depending on wind speed. Although the applications of on-grid systems are mainly in the residential sector, many commercial and industrial consumers have similar projects.

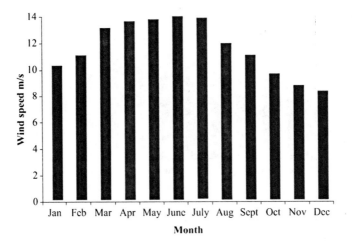

Figure 7.9. Monthly average wind speed at Alasfar site.

7.3.1 CASE STUDY 7: ON-GRID WIND SYSTEM WITH A 20 KW TURBINE IN JORDAN

7.3.1.1 Background

The continuous increase in the price of electricity in Jordan has affected the public and private sectors. The tendency to employ renewable energy for electricity generation was encouraged by the government, which has issued several laws to facilitate and adopt such projects. Although most remote areas in Jordan have access to the electric grid, the residents and users generally prefer to make use of the abundant renewable energy resources in these areas. One of these areas belongs to the Royal Jordanian Air Force (RJAF).

7.3.1.2 Site Characteristics

The considered site is called "Alasfar" hill in Almafraq province, 150 km (93 miles) to the east of Amman, the capital of Jordan. The site is located in a semidesert area at an altitude of 1,090 m (3,580 ft) above the sea level. The area surrounding the site is widely open with generally uninterrupted access to wind from all directions; there are no trees or buildings close to the wind turbine location. The hourly mean wind speed at the site is more than 11 m/s (25 mph) as shown in the monthly records of Figure 7.9.

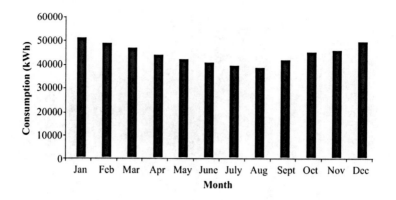

Figure 7.10. Monthly power consumption of Alasfar site.

Figure 7.11. PLC Module and touch screen for easy turbine control.

The wind is almost continuous with few windbreaks. Most of the strong wind periods were recorded at the end of spring and summer months, although winter winds are also known for their high speed. The calm time is very limited according to the long-term records provided by the Jordanian meteorological department.

7.3.1.3 Load Characteristics

The loads in this site are fed from the national grid, through a distribution transformer stepping down the voltage from 33 kV to 0.415 kV. The electricity consumption in winter is more than that in summer due to the heavy use of electric heaters. The place is known for its harsh coldness in winter months that extends from October to March. The average instantaneous load at the site is about 60 kW, while the peak can reach 70 kW.

Figure 7.12. Free-standing tower for 20 kW wind turbine.

The loads at the site can be divided into three main categories. The first category is the extremely vital load (Radar), which has a power rating of 40 kW, whereas the second category of loads is the water pumping system with a total rating of 6 kW. Finally, the remaining loads are estimated to be 14 kW. Since power continuity is an essential issue for this site, two diesel generators are kept running for several hours weekly to be ready for operation when the supply from the grid is interrupted. However, the cost of kWhs generated by such diesel generators are much more than that of power purchased from the grid. The recently increased prices of electricity have aggravated the energy problem in Jordan. Figure 7.10 shows the monthly consumption of this site.

7.3.1.4 Turbine Characteristics

The design of a selected wind energy system was based on the target to achieve the most reliable and efficient generation system. The adopted technology significantly reduced the weight and size of the generator down to 20%–30% of a more traditional one, and the production efficiency was increased by 10%–30%. This ensures that the generator functions perfectly in low wind conditions. The wind turbine controlling system uses a smart touch-screen Programmable Logic Controller (PLC) that is integrated with a Siemens controlling unit. The PLC will alarm and adjust any abnormal running status of the wind turbine automatically both timely and efficiently as shown in Figure 7.11.

The turbine is supported by an 18 m (59 ft) freestanding tower, made of fine, galvanized steel and strong enough to resist heavy wind. Figure 7.12 shows the free-standing tower during the turbine erection.

The rotor blades were designed to give the lowest noise possible. The design requirements were supported by advanced technology adopted by the manufacturer of this kind of wind blade. The result led to a clear reduction of noise to meet international standards. The turbine noise test has shown that it is within international standards limit. For example, at 30 m (98 ft) distance, the noise is less than 48 db.

A yawing system and an electromagnetic braking system are used to regulate the rotation of the wind turbine. In addition, a hydraulic braking system is a third protection. The present wind power system is designed to easily and conveniently shut down the whole system both manually and automatically in the event of excessive wind or major malfunction of the turbine unit.

One of the important characteristics of this turbine was the maximum power output guaranteed by the power curve as shown in Figure 7.13. To increase the efficiency of the generated power from the turbine, it is first converted from variable frequency power to high frequency power then to DC before it is supplied to the inverter. Other features, including the low cut-in and high cutoff wind speeds, are specified in Table 7.3.

7.3.1.5 Connection to Grid

The site is connected to the medium-voltage network through a 100 kVA 33kV/0.415 kV step-down transformer. According to the distribution util-

Table 7.3. Specifications of 20 kW wind turbine installed at Alasfarsite

Item	Value	Item	Value
Cut-in/cutoff wind speed, m/s (mph)	3/22 (6.7/49)	Generator type	Permanent magnet
Rated wind speed, m/s (mph)	11.5 (25.7)	Generator weight, kg (lb)	496 (1,090)
Generator efficiency	> 0.87	Blade material/quantity	GRP/3
Wind energy performance coefficient (C_p)	0.4	Overspeed control	Yawing + EM/hydraulic brake
Blade diameter, m (ft)	9.0 (30)	Shutting-down method	Manual + automatic

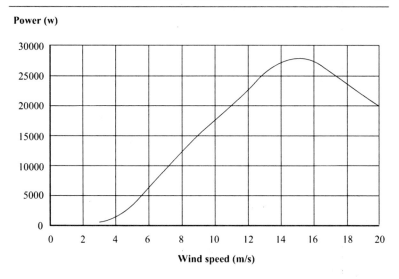

Figure 7.13. Power curve of 20 kW wind turbine installed at Alasfar site.

ity, responsible for the grid in Alasfar area, any connection to the utility network should be according to ENA G59/2 standards. This means that the installed inverters should be equipped with anti-islanding capability to protect the people working on the grid lines from back feed during power interruptions. Therefore, the inverter was selected to satisfy the required standards and local regulations concerning power and voltage qualities. The technology adopted in this inverter allows for a dual input section to process two strings with independent Maximum Power Point Tracking (MPPT) methodologies: high-speed and precise MPPT algorithms for real-time power tracking and energy harvesting. It was selected from the type of transformer-less operation for high performance efficiencies of up to 97%. The wide input voltage range makes the inverter suitable for low power installations with reduced string size. This outdoor inverter has been designed as a completely sealed unit to withstand the harshest environmental conditions.

7.3.1.6 Energy Production Assessment

The Jordanian tariff system considers this site as a public consumer, purchasing electricity at a fixed price of $0.16 per kWh. The total annual electricity bill for this site is around $71,000, not considering the fuel cost for the diesel generators used as a backup for emergency cases. This amount

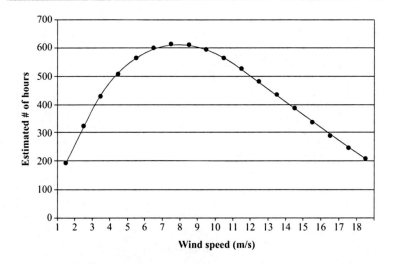

Figure 7.14. Probability distribution function for the wind at the considered site.

is usually subjected to an annual increase of around 20% due to the rise in fuel costs in the international market.

As the average wind speed in the site was known, a Weibull distribution was applied to describe the wind speed frequency curve using the probability density function. Figure 7.14 illustrates the estimated number of hours of wind speed at each range per day.

By using the wind-power equation and the estimated number of hours at each speed, it is possible to find the yield energy measured in kWhs. For the 20 kW wind turbine, different scenarios of operational time were considered and the payback period was calculated. In all considered cases, with various turbine prices taken from different manufacturers, the payback period was less than 5 years. This period will be shorter if we considered the continuous increase in electricity prices due to the frequent rise in fuel costs. However, the actual feasibility assessment of this project will not be clear until a sufficiently long operation is considered. Nevertheless, the first month records have shown that the turbine can produce more energy than expected.

7.3.2 CASE STUDY 8: ON-GRID SYSTEM WITH 80 KW WIND TURBINES

7.3.2.1 Background

In the late 1980s, the Royal Scientific Society (RSS) and the Ministry of Energy and Mineral Resources (MEMR) in Jordan established a wind resource assessment program. Through this program, 30 measuring stations were installed at sites throughout Jordan to map wind energy resources. The highest recorded annual average wind speeds were found in the northern part of the country (5.5–7.5 m/s [12–17 mph] at a height of 50 m [160 ft]). In 1989, MEMR and Meteorological Department, in cooperation with the Riso National Laboratory in Denmark, developed and published a preliminary wind atlas based on data collected from measuring stations around the country. It was based on data collected at 36 meteorological stations. All data analysis has been performed in support of the Wind Atlas Application Program (WAsP), in which the wind areas were identified for potential wind energy. In addition, it was found that the southern areas enjoy high wind speeds, while the eastern regions have significant wind potential [8].

7.3.2.2 Site Characteristics

The Al-Ibrahimiyah wind farm is a result of the efforts spent to activate the wind energy assessment program. This wind farm was established by the MEMR in cooperation with the RSS in April 1988 to be the first wind farm connected to an electric grid in Jordan. The farm is located in Irbid province, approximately 80 km (50 miles) to the north of Amman, the capital of Jordan. The altitude of the site is 1,030 m (3,400 ft) above sea level. The site has no natural obstacles or houses, hindering the wind from effective driving of the turbine. The wind blows smoothly with few periods of turbulence. The hourly mean wind speed in this site is 6.5 m/s (15 mph).

7.3.2.3 Turbine Characteristics

The farm consists of four wind turbines with a capacity of 80 kW each (320 kW in total). The turbines are of Danish design and manufacture. The hub height is 25 m (82 ft) above the ground, whereas the rotor diameter of the turbine is 17 m (56 ft). This corresponds to 253 m^2 (2,700 ft^2) of swept area. The annual production of the farm is 750 MWh of electricity, which is associated with a capacity factor of approximately 27% and availability of 96%. The turbines are stall regulated and the rated power of each turbine is reached at a wind speed of approximately 14 m/s (31 mph).

7.3.2.4 Connection to Grid

The wind farm is connected to the medium-voltage network through a set of step-up transformers. As the availability of the farm is high, the

Figure 7.15. A view of Al-Ibrahimiyah wind farm.

supply interruption is usually attributed to the grid performance and its reliability. In many cases, the national grid problems cause a significant loss of power generated from these turbines at good wind because there is no viable storage means. With the exception of minor faults, the farm has been operating successfully since it was built. The payback period of this project was slightly less than 7 years. Figure 7.15 shows a view of Al-Ibrahimiyah wind farm.

7.3.2.5 Public Acceptance

Local residents in the area have readily accepted the project, especially when they understood its role in reducing the pollution level and greenhouse gases. In general, this area is quiet and green, and since wind energy projects are environmentally friendly, there was no real objection to the installation of such projects.

7.3.3 CASE STUDY 9: ON-GRID WIND SYSTEM TO POWER-ISOLATED COMMUNITY, CANADA

7.3.3.1 Background

In 2004, Frontier Power Systems constructed a small wind farm near the community of Ramea, Newfoundland, Canada. The community of 700, a fishing village, is located on an island 10 km (6.3 miles) off the south coast of Newfoundland. Prior to the construction of the wind farm, all electricity was supplied by diesel generators, owned and operated by Newfoundland and Labrador Hydro. The wind farm produces 10%–13% of the 4,300 MWh consumed annually by the community, thus reducing the amount of fuel

Figure 7.16. Small wind farm near the community of Ramea, Canada.

purchased for the diesel generators. Newfoundland and Labrador Hydro pays Frontier Power Systems according to the diesel fuel savings realized by the wind farm. As an innovative renewable energy project, the Ramea system received financial assistance from Natural Resources Canada to test the new Wind-Diesel Integrated Control System (WDICS) [9].

7.3.3.2 System Description

The wind farm consists of six 65 kW wind turbines, a dump load, and an advanced automated control system. The turbines are reconditioned, stall-regulated machines mounted on lattice towers with a hub height of 25 m (82 ft) as shown in Figure 7.16.

Used turbines were employed in order to keep initial costs down. The controller is responsible for, among other functions, adjusting the power dissipated in a variable dump load. The load is adjustable in 1 kW increments between 0 kW and approximately 200 kW. The control system dumps wind farm output in order to keep the diesel generator operating at no lower than 30% of its rated capacity; below this loading, the generator operation is hazardous and unreliable, also the generator may wear prematurely. A sophisticated monitoring system monitors and logs the operation of the wind farm [9].

7.3.3.3 System Assessment

It was reported that the wind turbines do affect the operating regime of the diesel generators, but have little effect on the grid power quality (even during turbine connection). It was also shown that significant installation delays may be encountered due to regulations, bureaucracy, and uncooperative weather. The process of landlease from the Crown, and provincial and federal environmental assessments were also associated with some delay. Storms during winter impeded construction and low summer winds delayed commissioning [9].

The utility prepares an Operating Agreement as part of any Power Purchase Agreement (PPA), which defines the terms of the agreement between the utility and the nonutility generator. The operating agreement also defines the operating procedures and provides further clarification of the conditions of service outlined in the PPA. It states the minimum requirements for safe and effective parallel operation of the utility system with the generator facility and is intended for use by the generator and the utility when operating equipment, which will have an effect on the other party's systems [9].

This project has disclosed that wind–diesel systems are still a fairly novel technology, and problems are inevitable in early demonstration projects. The blade pitch setting of the turbines used in this project resulted in less than optimal performance and needed adjustment following commissioning. The use of reconditioned turbines, while decreasing costs, did impact reliability [9].

Even in an area known for strong winds, there can be significant variation in the wind regime over a fairly short distance. When the scale of a wind–diesel project does not justify an on-site wind resource measurement, and data from a somewhat distant weather station is used instead, the project becomes more risky. Newfoundland and Labrador Hydro (NHL) had already automated the diesel power plants by using relatively new diesel-generator sets that are equipped with advanced digital control and protection systems. The diesel plant automation straightened the integration process of the wind and the diesel generation and saved additional "interconnection" costs. NHL has a "sharing-the-savings" policy, which means, for example, if the wind project developer installs a project for $0.12/kWh and the avoided cost is $0.20/kWh then the PPA will be based on $0.16/kWh. Furthermore, if the plant efficiency is 4 kWh/L and the efficiency drops to 3.6 kWh/L when the wind plant is connected, the PPA will shift the costs to the wind developer. However, for the Ramea wind–diesel project, the total avoided cost will be paid to the wind developer, since it is the first Canadian project to demonstrate the wind–diesel technology. The purchase price details and policy are normally given as part of PPA between the utility company and the nonutility developer [9].

The use of refurbished wind turbines is critical in determining the economic feasibility of the project. The lower initial cost reduces the payback from 15 y to around 7 y. However, the availability of refurbished turbines and their cost may vary considerably.

There is substantial financial risk involved when embarking on remote wind projects. A number of factors can easily put an end to a seemingly profitable venture, particularly at the construction and commissioning phase. Access and weather are major considerations in planning and construction of these projects and should not be underestimated when evaluating the overall risk involved [9].

REFERENCES

[1]. Ackermann, T., & Söder, L. (2000). Wind energy technology and current status: A review, *Renewable and Sustainable Energy Reviews*, 4, 355–356.
[2]. Argaw, N., Foster, R., & Ellis, A. (2003). Renewable energy for water pumping applications in rural villages, *National Renewable Energy Laboratory, Subcontractor Report No. 500-30361*, 20–22.
[3]. Zietr, B. G. (2009). Electric wind pumping for meeting off-grid community water demands, *Guelph Engineering Journal*, 2, 14–23.
[4]. Bergeywindpower co. (2008). BWC Excel Wind Turbine, http://www.bergey.com/Products/Excel.html (accessed November 12, 2008).
[5]. Al Suleimani, Z., & Rao, N. (2000). Wind-powered electric water-pumping system installed in a remote location, *Applied Energy*, 65, 339–347.
[6]. American Wind Energy Association. (n.d.). Small wind, Case Studies, http://www.awea.org.
[7]. United Nations. (2001). Potential and prospects for renewable energy electricity generation in the ESCWA region, *A Report Prepared by United Nations*, New York.
[8]. Green, J. (1999). Small wind turbine application: Current practice in Colorado, *Colorado Renewable Energy Conference*, Winter Park, Colorado, September 10–12.
[9]. Natural Resources Canada, Power-wind turbine-390 kW-Isolated-grid/Canada, http://www.retscreen.net/ang/case_studies_390kw_isolated_grid_canada.php.

Index

A
Abraham, J.P., 5
Aerodynamic models, 49–50
Aerodynamic performance, 51
Akwa, J.V., 77, 78
Alaska, USA, 156–159
American Wind Energy Association (AWEA), 131
Applications
    case studies of small wind, 147–171
    off-grid small wind systems, 148–160
    on-grid wind systems, 160–171
    of small-scale wind power, 7–9
Aspect ratio, 74–75
AWEA. *See* American Wind Energy Association

B
Barakos, G.N., 143
Bazilevs, Y., 137
BEM theory. *See* Blade Element Momentum theory
Benesh, A.H., 80
Betz, A., 103
Betz's limit, 105
Beyene, A., 47
Bhutta, M.M.A., 47
Bianchini, A., 51
Blackwell, B.F., 79
Blade element geometry, 110
Blade Element Momentum (BEM) theory, 111–113, 134
British Wind Energy Association (BWEA), 131
Burton, T., 107
BWEA. *See* British Wind Energy Association

C
Cash flow model, for financial evaluation
    internal rate of return (IRR), 34
    net present value (NPV), 33–34
    payback periods (PBPs), 32–33
    return on investment (ROI), 33
    total cost of ownership (TCO), 34–35
Cellular communication tower, 4
CFD. *See* Computational Fluid Dynamics
Circulation control, 53
Clausen, P.D., 95
Coanda-type flow, 69, 70
Colorado, USA, 159
Computational Fluid Dynamics (CFD), 14, 49, 85, 119, 122, 129
Conventional Savonius-style turbines, 83
    coefficient of, 72
Costratio, 10
Crossflex concept, 53

D

Danish Energy Authority (DEA), 131
Darrieus and Savonius-style rotors, 2
Darrieus, George Jeans Mary, 45
Darrieus–Masgrowe-type wind turbines, 52–53
Darrieus-style turbines, 2, 3, 4
Darrieus-style VAWT, 6
Darrieus-type vertical-axis wind turbines
  aerodynamic models, 49–50
  with alternative airfoils, 51–52
  alternative vertical-axis wind turbine designs, 51
  computational methods, 50–51
  Darrieus–Masgrowe, 52–53
  design optimization, 49–53
  design parameters, 48–49
  experimental investigation, 50
  helically twisted blade, 52
  hybrid Darrieus–Savonius rotor, 52
  innovative vertical-axis wind turbine designs, 51
  orthogonal rotor, 52–53
Darrieus-type wind turbines
  current designs, 47–48
  practical darrieus vertical-axis wind turbine design methodology, 53–56
  vertical-axis wind turbines (VAWT), 48–53
DEA. *See* Danish Energy Authority
Design load cases and external conditions, 131–134
Design parameters, of Savonius-style turbines
  aspect ratio, 74–75
  blade profiles, 79–81
  end plates, 77
  multi-staging, 77–78
  number of blades, 78–79
  overlap ratio, 75–76
  scaling factor, 73
  separation gap, 76–77
  solidity factor, 73–74
Detached Eddy Simulation (DES), 140
Direct Numerical Simulation (DNS), 140
Dragging-type flow, 69, 70
Driving torque, 106
Dynamic stall, 51

E

Egg-beater turbine, 2
Electrical wind pump design, 152
End plates, 77
Eriksson, S., 47

F

FEA methods. *See* Finite Element Analysis methods
Fernando, M.S., 85
Financial considerations
  assembly and installation, 26–27
  cash flow model, 31–32
  costs of small-scale wind turbines, 25–28
  economic evaluation, 22–23
  economic impact, 23
  energy expenditure, reduction of, 28–30
  financial returns, 28–32
  government tax credits and subsidies, 31
  internal rate of return (IRR), 34
  maintenance, 27–28
  net present value (NPV), 33–34
  payback periods (PBPs), 32–33
  performance characteristics, 23
  power curve, 24–25

prevailing wind speed, 23–24
project evaluation, financial
  calculations, 32–35
rebates, 30
return on investment (ROI), 33
shipping cost, 26
system cost, 25
total cost of ownership (TCO),
  34–35
utility buyback arrangements,
  30–31
Finite Element Analysis (FEA)
  methods, 49
Flettner's rotor, 65
Flow analysis, 51
Flow field and power coefficient,
  51
Flow velocities of SB-VAWT, 55
Fluid Structural Interaction, 129
Force diagram, blade airfoil, 56
Free stream velocity, 104
Free stream wind flow, 69, 70

G
Glauert correction, 121
Glauert, H., 116
Gómez-Iradi, S., 143
Gorelov, D.N., 52, 53
Guide vanes, 53

H
Hangan, H., 95, 102, 120
Hau, E., 101
HAWTs. *See* Horizontal-axis wind
  turbines
Heelat Ar Rakah, Oman, 155–156
Helically twisted blade, 52
Helical-shaped Savonius-style
  vertical-axis wind
  turbines, 66
Himmelskamp, H., 114
Horizontal-axis wind turbines
  (HAWTs), 1
  angular momentum, 106–108

blade element momentum
  theory, 111–113
blade element theory, 108–111
blade shape, 118–120
with contoured blades, 8
control systems, 97
current designs and
  innovations, 97–99
linear momentum, 103–105
nacelle and yaw system, 95–96
performance, 122–125
power output prediction,
  99–100
prototype testing, 120–122
rotor, 95
rotor design, 117
rotor parameters, 117
small horizontal-axis wind
  turbines, 93–95
stall delay, 114–115
theory, 102–103
thrust coefficient correction,
  116–117
tip loss correction, 113–114
tower and foundation, 96–97
wind turbine design, 100–102
Hybrid Darrieus–Savonius rotor,
  52

I
IEC. *See* International
  Electrotechnical
  Commission
Internal rate of return (IRR), 34
International Electrotechnical
  Commission (IEC), 131,
  133
Islam, M., 47, 48, 49, 52, 56

J
JavaFoil prediction, 119, 120

K
Krivospitsky, V.P., 52, 53

L
Lanzafame, R., 124
Large Eddy Simulation (LES), 140
LES. *See* Large Eddy Simulation
Llasaria, Altiplano, Bolivia, 155

M
MacPhee, D., 47
Manwell, J.F., 101, 104, 117
MCS website. *See* Microgeneration Certificate Scheme website
MEMR. *See* Ministry of Energy and Mineral Resources
Microgeneration Certificate Scheme (MCS) website, 131
Ministry of Energy and Mineral Resources (MEMR), 166
Moa, J.O., 138
Modi, V.J., 85
Mohamed, M.H., 80
Mojola, O.O., 76
Munduate, X., 143

N
Naima water network, Morocco, 152–155
National Renewable Energy Laboratory (NREL), 95
Natural Science and Engineering Research Council (NSERC), 57
Net present value (NPV), 33–34
Net tangential and normal forces, 54
Newfoundland and Labrador Hydro (NHL), 170
NHL. *See* Newfoundland and Labrador Hydro
NPV. *See* Net present value
NREL. *See* National Renewable Energy Laboratory
NSERC. *See* Natural Science and Engineering Research Council
Numerical simulation, of HAWT
 boundary conditions, 139
 fluid-structure interaction, 141
 inputs to, 135–136
 meshing and node deployment, 136–139
 numerical solver, 139
 proper selection of solution domain, 136
 software, 141
 turbulence flow models, 139–141

O
Off-grid small wind systems
 annual energy output, 150–151
 background, 148
 cultural aspects, 151
 turbine specifications, 148–150
 for water pumping, 151–152
On-grid wind systems, 160–171
 Jordan, 161–166
 power-isolated community, Canada, 168–171
 Turbines, 166–168
Overlapping flow, 69, 70
Overlap ratio, 75–76

P
Payback periods (PBPs), 32–33
PBPs. *See* Payback periods
Plasma actuators, 53
Plourde, B.D., 4, 81
Power coefficient, 104
Power coefficient, of Savonius-style "S"-shaped wind turbines, 71
Power curve, 51
Power Purchase Agreement (PPA), 170

PPA. *See* Power Purchase Agreement
Practical Darrieus vertical-axis wind turbine, 53–56
Pressure coefficient, 143
Pressure distribution, 51
Product certification and code compliance, 131–134
Propeller HAWT, 7

R
Rangi, R.S., 45
RANS. *See* Reynolds-Averaged Navier-Stokes Simulation
Refan, M., 95, 102, 119, 120
Relative flow velocity, 54
Returning flow, 69, 70
Return on investment (ROI), 33
Revolutions per minute (RPM), 136
Reynolds-Averaged Navier-Stokes Simulation (RANS), 140
ROI. *See* Return on investment
Ronsten, G., 114
Rotor power coefficient, 104
Royal Scientific Society (RSS), 166
RPM. *See* Revolutions per minute
RR. *See* Internal rate of return
RSS. *See* Royal Scientific Society

S
Savonius-style rotors, 20
Savonius-style wind turbines
    design parameters of, 73–81
    dimensions of, 67
    mathematical and computational models, 86–87
    power coefficient of, 71
    testing and performance measures, 81–86
    torque coefficient of, 71
    working principle of, 69–72

SB-VAWT. *See* Straight-bladed Darrieus VAWT
Scaling factor, 73
Selig, M.S., 118
Separation flow, 69, 70
Separation gap, 76–77
SGS. *See* Sub-grid scales
Small-scale HAWT, 21
Small-scale Savonius VAWT rotor, 19
Small-scale VAWT, 22
Small-scale wind power
    advantages of, 9
    applications of, 7–9
    issues, installing wind power, 9–15
    overview, 1–6
Small-scale wind turbines
    case study, 35–43
    financial considerations, 22–35
    selection of, 20–22
Small wind turbines-HAWT style
    approach, 134–141
    power generation, 141–145
    product certification and code compliance, 130–134
    structural safety performance, 141–145
Snel, H., 114
Solidity factor, 73–74
S-shaped Savonius-style vertical-axis wind turbines, 66
Stagnation flow, 69, 70
Straight-bladed Darrieus VAWT (SB-VAWT), 47
Sub-grid scales (SGS), 140

T
Tangential velocity, 106
TCO. *See* Total cost of ownership
Teetering mechanism, 97
Telecommunication transceiver towers in Kenya, Africa
    background, 35

cash flow model, for financial
evaluation, 40
cost considerations, 37–39
financial calculations, 40–42
performance characteristics,
36–37
return considerations of, 39–40
Templin, R.J., 45
Thrust, 104
Thrust coefficient, 123
Tip speed ratio (TSR), 86, 105
Torque coefficient, 51
of Savonius-style "S"-shaped
wind turbines, 71
Total cost of ownership (TCO),
34–35
TSR. *See* Tip speed ratio
Twisted-bladed Savonius-style
vertical-axis wind
turbine, 82

U
UIUC. *See* University of Illinois at
Urbana-Champaign
University of Illinois at Urbana-
Champaign (UIUC), 118
Untaroiu, A., 50

V
VAWTs. *See* Vertical-axis wind
turbines

Velocity ratio, 54
Vertical-axis wind turbines
(VAWTs), 2. *See also*
Darrieus-type vertical-
axis wind turbines;
Practical Darrieus
vertical-axis wind
turbine
to cellular communication
tower, 4
connected to solar
panel, 10
Vironment, Aero, 98
Vortex flow, 69, 70
Vortex Wake method, 102

W
Wakui, T., 52
Walker, S.N., 117
WAsP. *See* Wind Atlas Application
Program
WDICS. *See* Wind-Diesel
Integrated Control
System
Wilson, R.E., 94, 117
Wind Atlas Application Program
(WAsP), 167
Wind-Diesel Integrated Control
System (WDICS), 169
Wind turbine types, 20–22
Wood, D.H., 95

THIS TITLE IS FROM OUR ENVIRONMENT ENGINEERING COLLECTION. OTHER COLLECTIONS INCLUDE...

Industrial Engineering
- Industrial, Systems, and Innovation Engineering — William R. Peterson, Collection Editor
- Manufacturing and Processes — Wayne Hung, Collection Editor
- Manufacturing Design
- General Engineering — Dr. John K. Estell and Dr. Kenneth J. Reid, Collection Editors

Electrical Engineering
- Electrical Power — Hemchandra M. Shertukde, Ph.D., P.E., Collection Editor
- Communications and Signal Processing — Orlando Baiocchi, Collection Editor
- Computer Engineering — Augustus (Gus) Kinzel Uht, PhD, PE, Collection Editor
- Electronic Circuits and Semiconductor Devices — Ashok Goel, Collection Editor

Civil and Environmental Engineering
- Environmental Engineering — Francis Hopcroft, Collection Editor
- Geotechnical Engineering — Dr. Hiroshan Hettiarachchi, Collection Editor
- Transportation Engineering — Dr. Bryan Katz, Collection Editor
- Sustainable Systems Engineering — Dr. Mohammad Noori, Collection Editor

Material Science
- Materials Characterization and Analysis — Dr. Richard Brundle, Collection Editor
- Mechanics & Properties of Materials
- Computational Materials Science
- Biomaterials

Not only is Momentum Press actively seeking collection editors for Collections, but the editors are also looking for authors. For more information about becoming an MP author, please go to http://www.momentumpress.net/contact!

Announcing Digital Content Crafted by Librarians

Momentum Press offers digital content as authoritative treatments of advanced engineering topics, by leaders in their fields. Hosted on ebrary, MP provides practitioners, researchers, faculty and students in engineering, science and industry with innovative electronic content in sensors and controls engineering, advanced energy engineering, manufacturing, and materials science. **Momentum Press offers library-friendly terms:**

- perpetual access for a one-time fee
- no subscriptions or access fees required
- unlimited concurrent usage permitted
- downloadable PDFs provided
- free MARC records included
- free trials

The **Momentum Press** digital library is very affordable, with no obligation to buy in future years.

For more information, please visit **www.momentumpress.net/library** or to set up a trial in the US, please contact **mpsales@globalepress.com**.

CPSIA information can be obtained
at www.ICGtesting.com
Printed in the USA
FFOW05n0245150714